Diatom Stratigraphy and Human Settlement in Minnesota

J. Platt Bradbury
Limnological Research Center
Department of Geology and Geophysics
University of Minnesota
Minneapolis, Minnesota 55455

THE
GEOLOGICAL SOCIETY
OF AMERICA

SPECIAL PAPER 171

Library of Congress Catalog Card Number 75-21066
I.S.B.N. 0-8137-2171-7

Published by
THE GEOLOGICAL SOCIETY OF AMERICA, INC.
3300 Penrose Place
Boulder, Colorado 80301

The Special Paper Series was originally made possible
through the bequest of
Richard Alexander Fullerton Penrose, Jr.

Printed in The United States of America

Contents

TABLE

PLATE SECTION

Acknowledgments

This compilation has become possible only with the able assistance of several individuals who have been directly and indirectly involved in the stratigraphic analysis of diatoms from lakes in this region during the past two years. I am particularly indebted to Nancy P. Sather for her diatom analyses from the sediments of two bays in Lake Minnetonka and for her additional work on the Pickerel Lake core, and to Donna M. Stark for her analyses from Sallie and St. Clair Lakes. Kathleen Baker provided the analyses from Tanager Lake and identified the diatoms from the surface waters of Lake Minnetonka for several summers. Stark, Sather, E. B. Robertson, and G. L. Jacobson provided pollen analyses for several cores, principally under the tutelage of Jean C. B. Waddington. These and other persons have not only provided technical assistance but have enthusiastically discussed the interpretation of the results and have contributed in many casual ways to the development of the ideas and information presented here. Support for this study was provided by the Atomic Energy Commission (contract AT (11-1)-2046), the National Science Foundation (Grant GB40200), the U.S. Department of the Interior Office of Water Resources Research, the U.S. Environmental Protection Agency, and the Minnesota Resources Commission.

Abstract

Fossil diatom assemblages in short cores of lake sediment from nine lacustrine environments (seven lakes) in Minnesota and South Dakota show the ecological reaction of freshwater diatoms to limnologic changes associated with enrichment following European-American settlement. The time of settlement is stratigraphically determined by an increase in the proportion of *Ambrosia* (ragweed) pollen, which signals late 19th century land clearance and cultivation in this region. In some cases other stratigraphic profiles, such as phosphorus or mining wastes, are used to date settlement activities near the lakes. Marked changes in the diatom stratigraphy frequently correlate with the time of settlement around the lakes, and in most cases it has been possible to interpret these changes in the context of limnologic modifications caused directly or indirectly by the settlement activities of man.

Initially, diatom diversity decreases as the lakes become enriched by increased erosion and (or) by disposal of municipal wastes. Littoral (epiphytic and benthic) diatoms become underrepresented in the sedimentary record compared to predisturbance times, perhaps because of excessive shading by blue-green algae, or simply because they are numerically swamped out by massive blooms of planktonic diatoms. As enrichment increases, the more or less even distribution of spring, summer, and fall planktonic diatoms changes to a planktonic diatom flora dominated by species that bloom in the early spring—sometimes even under the ice. Foremost among these are *Stephanodiscus minutus* and *S. hantzschii*, the latter characterizing the most eutrophic lakes studied. Apparently the summer and fall diatom plankton cannot compete with the massive blooms of floating, blue-green algae that occur in the warmer seasons. Only in the shallow turbulent lakes do the heavy summer and fall diatom plankters maintain sizable populations that effectively compete with the buoyant blue-green algae. The stratigraphic record of the blue-green algae is inferred by the stratigraphy of *Chydorus sphaericus*, a normally littoral cladoceran that utilizes blue-green algal filaments to suspend itself in the limnetic regions of a lake, thus vastly increasing its habitat and population size.

There are several variations on this theme, depending on the initial trophic state of the lake and other limnologic characteristics such as basin morphometry, but overall, the diatom stratigraphy of lake sediment is an effective way to assess man's impact on the lake ecosystem.

Introduction

Recent interest in environmental quality has created a need for specific objective measures of the change in water quality of lakes and ponds since European-Americans settled in their watersheds and along their shores. Stratigraphic diatom studies serve the purpose. Diatoms are sensitive to changes in the chemical and physical properties of lakes, and their resistant siliceous shells commonly allow them to be preserved as fossils in lake sediments. Thus the stratigraphic record of diatom fossils can document the history of limnologic changes. Independent stratigraphic evidence signaling the arrival of settlers in the area frequently coincides with these diatom changes, which therefore are assumed to have been caused by man.

This information gives us an objective view of precultural and postcultural limnology that is site-specific (that is, it refers directly to the unique limnologic conditions of a single site), and it may be a valuable aid in directing and evaluating attempts to restore lakes to predisturbance conditions. As information from many sites accumulates, and as the general responses of diatoms and other components of lake ecosystems to cultural modification become known, it will be possible to use diatoms to monitor water-quality changes over larger areas. Undesirable changes in a lake could then be detected and arrested before major harm is done.

The success of this approach depends primarily upon two questions: (1) What is the ecologic significance of the stratigraphic diatom changes, and (2) Were they caused by man? The first question relates to the amount and accuracy of ecologic knowledge available about species and communities of freshwater diatoms. The existing knowledge has been gathered mainly from distributional studies of living diatoms from a wide variety of chemically and physically studied aquatic environments and by field and laboratory experiments designed to assess the effect of chemically and physically altered environments on the composition and growth rate of the resident diatom community. Despite the progress made in these areas, it is seldom possible to define the precise relationship of a diatom community, or even of a single species, to its environment. The complex of interacting factors to which algae respond in natural aquatic environments is rarely duplicated in laboratory experiments, and distributional studies generally do not measure these interactions. These problems limit the accuracy of paleolimnologic interpretations based on diatom stratigraphy. The matter is further complicated by the fact that the fossil assemblage upon which the interpretations are based may have no living counterpart at a given time and place in a lake, and therefore existing ecologic knowledge can only be applied to some fraction of the fossil assemblage.

The second question involves testing the assertion that stratigraphic changes in the diatom assemblages coincident with man's tenure around a lake are causally related to man. There may be no way of directly proving this, but the assumption is strengthened when a long presettlement record of diatom assemblages fails to replicate the postcultural changes. The strength of the stratigraphic approach is that it can easily provide the historical perspective necessary to evaluate suspected man-caused changes.

This paper summarizes the recent stratigraphic history of diatoms from seven lakes in Minnesota and South Dakota (Fig. 1) and documents the changes that can be ascribed to man's direct or indirect alteration of the limnologic environment.

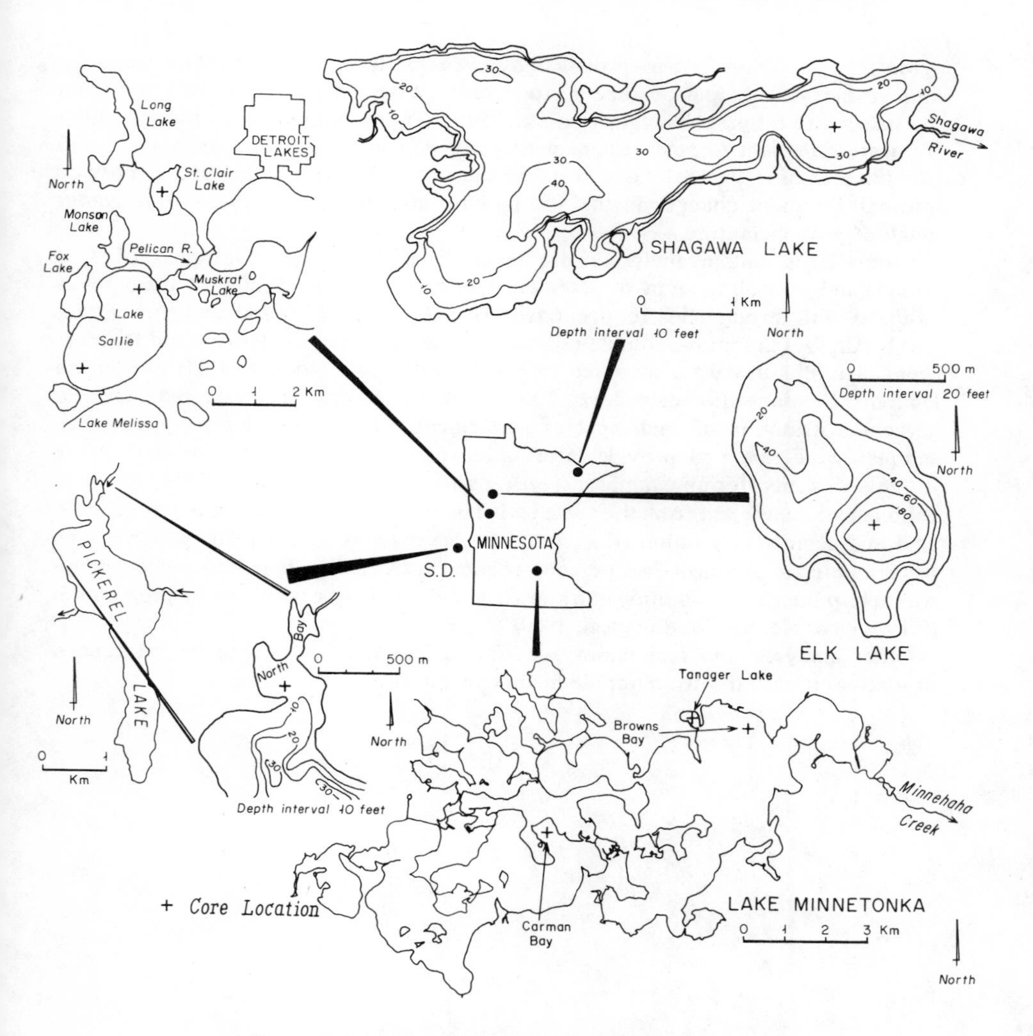

Figure 1. Index map of lakes and core locations.

Methods

Material for analysis was provided by cores of lake sediment taken from the lake surface with a plastic tube piston corer (Wright and others, 1965) or from the lake bottom by scuba. Short cores (60 to 160 cm long) covered the transition from presettlement to postsettlement time. In Pickerel Lake (Haworth, 1972) and Elk Lake (Stark, 1971), long cores, 750 cm and 1,200 cm, respectively, provide information about changes in the stratigraphy of diatom assemblages throughout much of postglacial time.

Samples for diatom analysis were taken from the cores at intervals of 2 to 10 cm, and a small amount of sediment (generally between 0.25 and 1 cm^3) was oxidized with strong acids (concentrated H_2SO_4 and HNO_3) and oxidants (H_2O_2 and $K_2Cl_2O_7$) to remove soluble minerals and organic matter (Patrick and Reimer, 1966). The diatoms were mounted in Hyrax (refractive index = 1.65) or Clairite Diatom Mountant. For each level 400 to 1,000 frustules were counted, and the percent distribution of each species was calculated. In some cases quantitative samples were taken to provide information on diatom concentration and influx to the sediments (for an example, see Bradbury and Waddington, 1973).

Pollen was extracted from the same levels by sediment digestion with hydrofluoric acid and acetolysis solution (Faegri and Iversen, 1964). Usually 200 pollen grains were counted to establish their percent frequency. Occasionally pollen concentration was determined by the addition of a known quantity of exotic pollen to the preparation (for an example, see Waddington, 1969).

Other analyses and techniques specific to individual sites will be referenced or discussed with the stratigraphic history of the site.

Nature of the Data

To evaluate the stratigraphic changes of diatom assemblages in terms of paleoecology, it is first necessary to discuss some of the interrelations between diatom ecology, sedimentation, and preservation and to show how these factors affect the primary data of the micropaleontologist. In a sense, this is an exploration of the assumption that the stratigraphic record of diatom fossils can document the history of limnologic changes. It is not my intent to prove or disprove this assumption, but simply to discuss briefly the nature of the data and the interpretations that might issue from them.

There are five principal factors that make diatoms important microfossils for studying the past history of aquatic environments.

1. Abundance. In freshwater lakes diatoms often compose the dominant algal group. Densities of planktonic associations in Minnesota lakes vary throughout the year and from lake to lake. Low diatom abundance is about 5 to 20 cells per millilitre, and high abundance is from 500 to 1,000 cells per millilitre, for samples collected in the fall (Tarapchak, 1972). During the spring diatom bloom in eutrophic Lake Minnetonka, there were 4,000 cells per millilitre (Megard, 1972). Attached-diatom communities are also large. Community densities of species attached on glass slides from oligotrophic lakes in Ontario range from 18,000 cells per square centimetre to more than 330,000 cells per square centimetre, and on natural substrates the density may exceed 1,000,000 diatoms per square centimetre. This figure includes an unknown number of dead individuals within and beneath the living community (Stockner and Armstrong, 1971). In Lake Winnipeg, which is more eutrophic, densities of more than $6{,}000{,}000/cm^2$ have been recorded (Evans and Stockner, 1972).

These figures indicate that there is an ample diatom biomass in freshwater lakes, and they suggest that diatoms could be a common constituent of lake sediments. Observation bears this out. Some lake sediments (diatomites) are almost exclusively composed of diatoms, and in Shagawa Lake, northeastern Minnesota, diatoms form more than 30 percent of the dry sediment weight and have a concentration of more than 2×10^8 cells per cubic centimetre.

2. Ecologic diversity. Diatoms appear to be ecologically diverse. There are probably about 1,000 common species and varieties of freshwater diatoms in north temperate Europe and America. They occur not only in the plankton but in a variety of benthic habitats, where they are either attached or motile. Attached species grow on plants (epiphytic), rocks (epilithic), mud (epipelic), and many other substrates, both natural and artificial. They may be commensal with other organisms, or they

may live among the sand grains of wave-washed beaches. Some attached species are found predominantly in aerial habitats, for example, on tree bark or on damp rocks. Despite the fact that a single species may live in more than one habitat, there are enough differences so that each habitat has a characteristic community. In addition to the physical habitats, there are several other environmental factors that are important in the distribution of freshwater diatoms. Light, temperature, salinity, pH (alkalinity), nutrient level, and current are important, as are the amounts of many elements and compounds dissolved in lake water. Because of this great ecologic diversity and sensitivity, diatoms have a great potential for paleoecology.

3. Preservation. Diatoms possess shells (frustules) of silica, which resist destruction by oxidation. As a result, they are commonly preserved in lake sediments after other algae have been destroyed.

4. Identification. The taxonomy of diatoms is based on frustule characteristics; thus the fossils are quite as identifiable as the living species. In fact, precise identification of living species can seldom be accomplished until the cellular contents are removed and the frustule is mounted so that its morphology can be easily seen.

5. Quantification. Generally the fossil remains of diatoms can be easily related to the individual organisms that produced them, and a count of the remains therefore provides direct information about population and community density. The frustules may be preserved entirely, or they may separate into the two component valves. In the latter case, each valve counts for half an organism. Even broken valves, if they can be identified, can be related back to the individual diatom, since each valve has some distinctive morphological element or elements that can be quantified, such as the central nodule or terminal nodules. Because of this, fragments of one diatom will not be counted as representing more than one individual.

Despite the optimistic tenor of the foregoing, diatoms are not independent of the sedimentary environment that controls their ultimate deposition and preservation, and this close relationship introduces many variables that complicate paleoecologic interpretations of diatom assemblages. Sedimentation rates in lakes depend on lake morphometry and drainage area, climate, vegetation, and many other factors. Fast sedimentation rates may dilute the accumulation of diatoms, but at the same time, rapid burial may prevent many from being destroyed by bottom-feeding organisms or by solution, especially in alkaline lakes. The converse is true of slow sedimentation rates. The time it takes for a given thickness of sediment to accumulate permanently (deposition time) is a reflection of the sedimentation rate and the energy of the sedimentary environment, which control where and how much sediment moves around on the bottom before it becomes buried. Currents in lakes may be associated with the inflow and outflow of rivers, turnover of the water column during spring and fall, and exposure of the lake to wind. Such water movements can differentially break diatoms, winnow the lightweight ones away from heavier, larger ones, or provide environmental conditions that favor their ingestion by benthic scavengers. At the same time, such currents bring together diatoms from many different habitats of the lake. The accumulating assemblage is probably not in a constant proportion to the areal extent of the habitat, although the closest and most productive habitats contribute the most diatoms (Merilainen, 1971).

wind. Until the local and regional components of the *Ambrosia* rise can be separated, this stratigraphic marker can give us only a general idea of the chronology of postsettlement disturbance.

Another way to approach the problem is to look more closely at the other components of the *Ambrosia* rise, such as weed species introduced into North America at a known date. A potential candidate for this approach is *Salsola kali* (= *Salsola pestifer*, Russian thistle). This chenopodiaceous weed was introduced into South Dakota in 1874 with imported flax seed. By 1888 it was common enough in the Dakotas to be recognized as a weed, and ten years later it had spread throughout the area east of the Rocky Mountains from the Gulf of Mexico to Saskatchewan (Weaver and Clements, 1929). The appearance of *S. kali* in pollen diagrams can thus provide a maximum age for that horizon, depending on the geographic location of the profile. The potential of this approach can be realized only if large numbers of pollen grains are counted to assure a statistically reliable absence of the pollen type. Additional refinements could be made if the migration history of *S. kali* and its capacity to disperse pollen were more perfectly known.

Other chronologic markers exist as special cases in some kinds of lake basins. Where annual layers of sediment are well-defined, a precise chronology is available if the layers extend to the surface or to some stratigraphic marker of known age. Such stratigraphic markers are often unique man-caused sedimentary events, like the appearance of hematite (iron ore) particles in the sediments of Shagawa Lake as a result of iron mining starting in 1890. Similarly, abrupt increases or first appearances of phosphorus, lead, mercury, DDT, or other pollutants in the profiles may provide a chronology if the history of their use is known for that area.

Man's arrival and intensive settlement in an area may involve the destruction of the natural vegetation, which in turn causes an increase in surface runoff and erosion of topsoil. Nearby lakes can stratigraphically record these events as an increase in the concentration of silt and clay in the profile (for an example, see Davis, 1970). Similarly, pollen and terrestrial fungi that are retained in topsoil litter may be washed into these lakes and deposited when erosion increases. In such cases there may be an abrupt increase in the pollen concentration in the sediment. Assuming that the regional pollen rain is more or less constant over short time periods, these increases in pollen concentration may be interpreted as indicating the time when man's agricultural activities directly affected a given lake (Bradbury and Waddington, 1973). In this respect pollen concentration increases may reflect local conditions more precisely than the regional chronology provided by the *Ambrosia* rise. Not all sediments of disturbed lakes show increased proportions of material eroded from the surrounding landscape, however, because the same influx of mineral and organic matter may increase the productivity of the lake, and the resulting increase in organic sedimentation can mask the increase in allochthonous material. A separate measure of influx per year can be calculated, however, if the chronology is known.

Diatom Ecology

The distribution of diatoms has attracted the attention of phycologists almost since the beginning of diatom studies more than 100 years ago. The early efforts were confined principally to geographic comparison of diatom floras. Little attempt was made to seek the ecologic or physiologic reasons underlying the floristic differences, although the obvious characteristics of marine and freshwater habitats were recognized. The distributional classification of diatoms grew rapidly but casually, so that by 1900 diatomists had looked for and found diatoms in most conceivable places, and both the marine and nonmarine habitats were extensively subdivided.

The marine habitats were initially divided on the basis of geography and were later subdivided according to the location and influence of current systems. Generally, the distributional patterns were based on the species encountered. In contrast, diatomists working in freshwater areas imposed a popular classification of aquatic environments (that is, ponds, streams, bogs, lakes, waterfalls, and so forth) on the organisms, and not surprisingly, many species were found to overlap in their "ecological" distribution.

By the end of the 19th century, limnologists had begun to categorize aquatic environments on the basis of the chemical, physical, and biologic characteristics of the water. European phycologists soon incorporated water analyses with the study of freshwater diatoms and were able to define communities and indicator species that had reasonable consistency. Thus the halobion spectrum (Kolbe, 1927), the pH spectrum (Hustedt, 1937-39), and others were defined (Kolbe, 1932). Subsequent investigations in many other parts of the world, however, have shown that the ecologic and geographic distribution of many diatom species is extraordinarily wide. Cholnoky (1968) discussed the sources of error and inadequacies of the traditional concepts of diatom ecology and illustrated with case studies how diatomists adapted to the influx of additional information by ever subdividing the ecologic categories or by describing the ecologic-geographic distribution of diatoms in ever more general terms.

To some extent the inadequacies result from the fact that modern investigators are asking more sophisticated ecologic questions than did their predecessors—questions aimed principally at understanding the physiologic responses of different species to varying water conditions. As this information accumulates, there is a natural tendency to play down the earlier efforts. At the present time, there is nothing to replace the earlier artificial system, however simplistic it may be, and many diatom ecologists are faced with the unhappy situation of using a possibly outmoded system of ecologic characterization or refraining from making any ecologic assertions and generalizations at all.

Actually the situation is not so grave as it might at first appear. It is true that autecologic information about any given species is largely unavailable, and as a consequence it is nearly impossible to say anything specific about the ecologic significance of the presence or absence of a single diatom "indicator" species. Associations of diatoms can be used with greater success, however, even though the ecologic inferences from associations must be more general, partly because ecologic information has not been conveniently summarized and partly because the differences in both kind and frequency of species that usually exist between similar associations make exact comparison difficult. At present there are no guidelines on how to evaluate such differences.

Paleoecologic inferences from fossil diatom assemblages are further generalized because short-term variations in the diatom community are averaged in the sedimentary record. Because the site from which a core is taken for paleoecologic study can generally be assumed to have had a reasonably constant relationship to the morphometry of the basin and to the pattern of sedimentation, it is usually clear that when large stratigraphic variations are seen in the distribution of fossil diatoms, these record real limnologic changes. It is also evident, however, that inferences from such data are less specific because fossil assemblages are seldom replicated by modern associations.

The common diatoms occurring in Minnesota lake deposits can be grouped broadly into two major habitat types: planktonic and benthic. A graphic representation (Fig. 2) of habitat data for some common Minnesota diatom species (Bright, 1968) shows that the distinction is not always clear-cut and that some species that are commonly considered planktonic, such as *Tabellaria fenestrata* and *Fragilaria capucina* var. *mesolepta*, are present in considerable numbers in benthic habitats. Although these data suggest that the habitat distribution for many species may be broad, it must be noted that the collections are not completely comparable, because samples from benthic habitats often are unaccompanied by correlative plankton samples. Even if they were, both benthic and planktonic diatom communities exhibit great temporal fluctuations in species composition and size, and it is unlikely that the absence of a species at one sampling time means that it never occupies that habitat. This is a case where the diatom assemblages of surficial sediments would provide a useful measure of diatom productivity and composition in different habitats. Merilainen (1971) has studied the distribution of diatoms in surficial sediments in meromictic lakes and has shown a clear separation of benthic and planktonic floras.

The distribution of diatoms in Minnesota lakes according to lake chemistry (particularly salinity) can also be seen from a graphic representation of Bright's (1968) chemical data (Figs. 3, 4). Salinity increases from the pure, low-salinity lakes in northeastern Minnesota to the comparatively saline and alkaline lakes in the west and southwest. In Figures 3 and 4, salinity is represented by conductivity. It is evident that for both planktonic diatoms (Fig. 3) and benthic and meroplanktonic (temporarily planktonic) diatoms (Fig. 4), some species are more or less restricted to lake water of low conductivity, others have a wide distribution, and some are restricted to lakes with high conductivity. The uneven data distribution in Figures

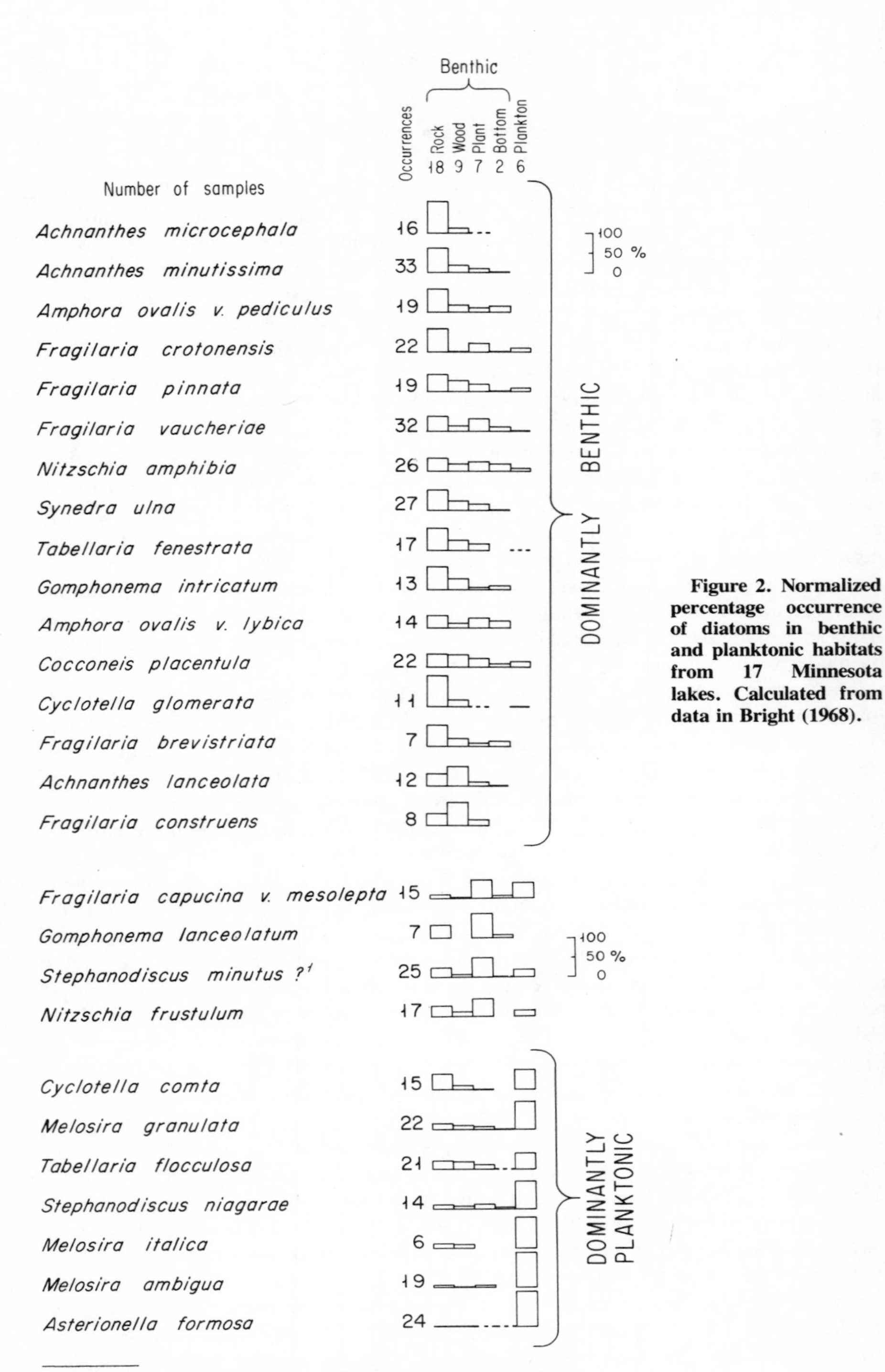

Figure 2. Normalized percentage occurrence of diatoms in benthic and planktonic habitats from 17 Minnesota lakes. Calculated from data in Bright (1968).

[1] Originally reported as *Stephanodiscus astraea* var. *minutula*.

3 and 4 probably results from the inability of a single sample to represent adequately a lake's diatom flora and also from the fact that many other variables besides salinity affect diatom distribution. In view of these limitations, the mere suggestion of a distributional pattern argues for the importance of lake salinity as a controlling factor in diatom distribution.

The productivity of Minnesota lakes is a subject of considerable interest, but so far there have been few regional studies of algal nutrients in natural lakes. Moyle (1954) showed that total phosphorus concentrations of lake water increased from low values in the northeast to higher values in the southwestern part of the state. This gradient generally follows the salinity gradient, although the two are not well correlated and not necessarily related, especially in disturbed lakes. Nevertheless, the diatoms prevalent in the moderate- to high-conductivity lakes (Figs. 3, 4) are those that commonly occur in eutrophic lakes.

Perhaps the best survey of planktonic diatoms with respect to nutrient enrichment comes from the studies in Lake Michigan, where Stoermer and Yang (1970) reviewed algal collections from the lake that were taken nearly 100 years ago and integrated these data with many more recent collections from polluted and unpolluted areas in this lake. They were able to identify dominant diatoms characteristic of highly oligotrophic environments, moderately oligotrophic offshore environments, and

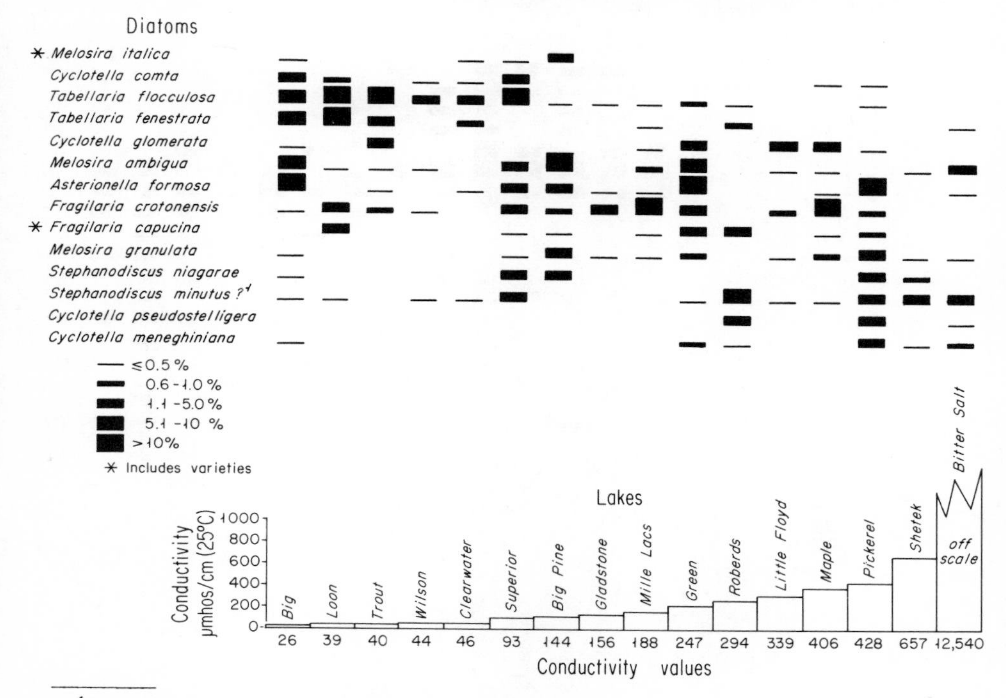

[1] Originally reported as *Stephanodiscus astraea* var. *minutula*.

Figure 3. Distribution of common "planktonic" diatoms from Minnesota lakes according to increasing water conductivity. Data from Bright (1968).

eutrophic "small lake" environments. In addition they found a number of widespread or "eurytopic" dominants and a few forms without clear distributional patterns. Whereas some of the diatoms are largely restricted to the Great Lakes, many of the forms are common also in Minnesota lakes, and the ecologic generalizations from the Lake Michigan flora are quite useful for interpreting past and present trophic conditions of lakes in Minnesota. The trophic relationships of the following diatoms are summarized from Stoermer and Yang's (1970) results:

Highly Oligotrophic Environments
- *Cyclotella kutzingiana*
- *C. ocellata*
- *C. operculata*
- *C. comta* (possibly of eurytopic distribution)

Oligotrophic Offshore Environments
- *Melosira italica* subsp. *subarctica*
- *Synedra ulna* var. *chaseana*
- *S. delicatissima* var. *angustissima*
- *S. filiformis*
- *Rhizosolenia eriensis*
- *Cyclotella michiganiana*

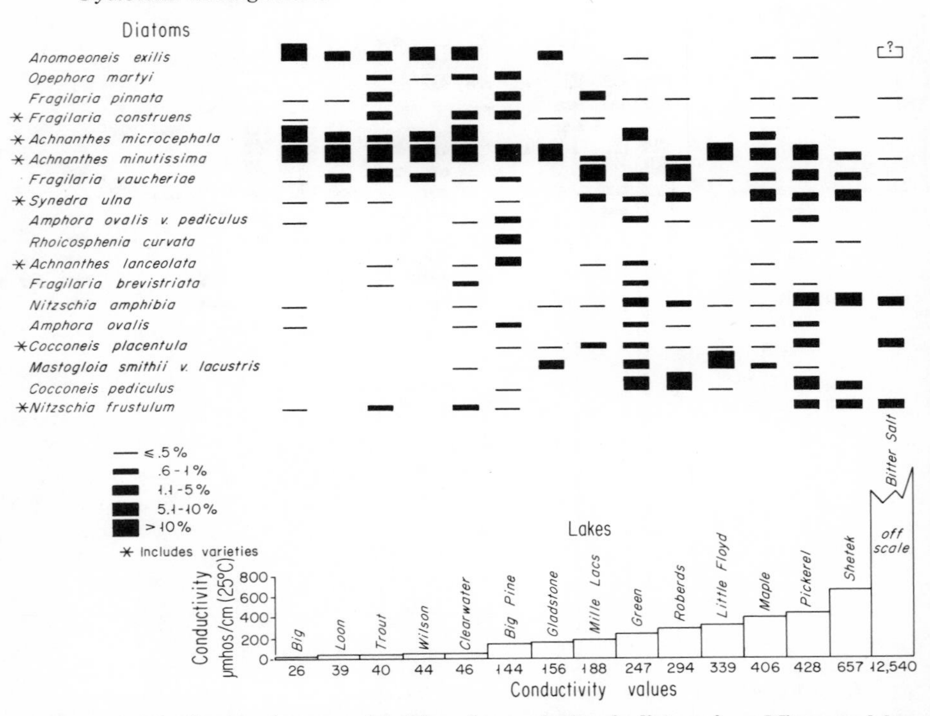

Figure 4. Distribution of common benthic and meroplanktonic diatoms from Minnesota lakes according to increasing water conductivity. Data from Bright (1968).

Eutrophic Small-Lake Environments

- *Fragilaria capucina* and *F. c.* var. *mesolepta*
- *Melosira granulata* and *M. g.* var. *angustissima*
- *Stephanodiscus binderanus*
- *S. hantzschii* (and several other small *Stephanodiscus* species)
- *Cyclotella pseudostelligera*
- *C. meneghiniana*

Eurytopic Diatoms

- *Tabellaria fenestrata*
- *T. flocculosa*
- *Asterionella formosa*
- *Melosira islandica*
- *Fragilaria crotonensis*
- *Stephanodiscus minutus*
- *S. niagarae*
- *Synedra ulna* var. *danica*

Stoermer and Yang (1970) did not report *Melosira ambigua* in their study of Lake Michigan phytoplankton, but surveys by Holland (1968, 1969) indicate that this species is present and characterizes eutrophic conditions along with *M. granulata*.

The most common planktonic diatoms in the smaller, generally oligotrophic lakes of the Canadian Shield are *Synedra acus* var. *radians*, *S. a.* var. *angustissima*, *Asterionella formosa*, *Cyclotella comta*, *C. stelligera*, *Tabellaria fenestrata*, *T. flocculosa*, *Rhizosolenia longiseta*, and *R. eriensis* (Schindler and Holmgren, 1971).

Many of the oligotrophic dominants of Stoermer and Yang's (1970) listing are not common enough in Minnesota lakes to be significantly represented in Bright (1968), and it is likely that for the most part the smaller Minnesota lakes are more productive than the undisturbed parts of Lake Michigan. The higher ratio of littoral area to total lake volume may be an important factor in determining the productivity of a lake, and this may account for the observed differences in the diatom floras. In contrast, the diatom dominants of the Canadian Shield lakes are also common Minnesota associations, particularly in the low-salinity, less productive lakes.

The benthic diatom communities of Minnesota lakes have received much less attention than the planktonic community. Bright's (1968) survey of Minnesota diatoms includes their littoral habitats, and this perhaps represents the most relevant information. Nevertheless, there is considerable scatter in the distribution of many species (Figs. 2, 4). Characterization of lake chemistry and productivity on the basis of benthic diatoms is complicated because the variables of this habitat include not only the composition of the lake water but also the composition and structure of the substrate. The interaction of these variables with differences in turbulence, light, temperature, and so forth probably account for the marked spatial and temporal changes of littoral diatom communities.

Plant and lake-bottom substrates should be a good source of organic and inorganic nutrients for diatoms. Jorgensen (1957) has shown that silica in *Phragmites* stems is easily soluble and suggests that epiphytic diatoms use *Phragmites* as a silica source when they bloom during the spring and winter. Probably other nutrients such as phosphorus and nitrogen are similarly available to benthic diatoms. Even species traditionally considered planktonic may take advantage of the nutrients in the littoral environment during certain times of the year. This is particularly true in oligotrophic lakes, where nutrient-poor water does not support large planktonic communities in the limnetic zone. Stockner (1971) and Stockner and Armstrong (1971) described this situation in the oligotrophic lakes of the Canadian Shield. Here, species of the usually planktonic genera *Melosira, Tabellaria,* and *Cyclotella* are common inhabitants of the littoral area during the spring. In some cases these diatoms become planktonic during the summer when turbulence is greater, but the productivity of the limnetic region is generally much lower than that of the littoral.

Similar habitats in large lakes may also be complex. The attached diatoms on navigational buoys in Lake Winnipeg, which would seem to represent a comparatively well mixed, uniform environment, were shown by Evans and Stockner (1972) to vary considerably in kind and number, apparently in response to nutrients derived from river plumes entering the lake and turbulence from lake currents and seiches. This is in contrast to the periphyton (principally attached diatoms) of Lake Superior. Fox and others (1969) showed that the species composition of the Lake Superior flora is constant along a 106-mi stretch of the north shore. *Achnanthes microcephala* and *Synedra acus* are the dominant species of this oligotrophic environment. The occurrence of *A. microcephala* is consistent with the Minnesota small-lake data (Fig. 4), which show that it is also abundant in the low-salinity lakes.

The benthic diatom flora of the Canadian Shield lake no. 240 is dominated by *Achnanthes minutissima* and *Fragilaria pinnata,* and the following types are common at least seasonally (Stockner and Armstrong, 1971): *Cymbella tumidula, Navicula radiosa* (*), *Cyclotella comta, Cymbella frigida, Tabellaria flocculosa, Anomoeoneis serians, Cymbella ventricosa* (*), *Gomphonema angustatum* (*), *Fragilaria crotonensis* (*), *Cymbella microcephala* (*), and *Anomoeoneis vitrea*. Lake no. 240 is regarded by Schindler and Holmgren (1971) as mesotrophic (midsummer C^{14} production ≤ 0.5 g $C/m^2 \cdot$ day). The productivity of 18 lakes from northeastern, northwestern, and southern Minnesota is mostly higher than this, with the most productive ones located in the south (0.8 to 5 g $C/m^2 \cdot$ day) and the least productive in the northeast (0.4 to 2 g $C/m^2 \cdot$ day) (Megard, 1970). Nevertheless, the starred diatoms (*) in the above list were shown by Bright (1968) to be widely distributed throughout Minnesota. For these species, and probably many other benthic diatoms, lake productivity is not an important factor except indirectly through its influence on macrophytes and turbidity.

The ecologic information presented on the diatoms of Minnesota and adjacent areas is no more sophisticated than the classical approaches to diatom ecology that deal with "spectra" of alkalinity, salinity, and so forth. It has the clear advantage of being geographically relevant to the area studied, but until we are able to obtain more precise and specific information about the distribution and abundance of the common species, one can only draw general conclusions about paleolimnologic changes from diatom stratigraphy. By combining diatom stratigraphy with other stratigraphic approaches such as geochemistry, pollen analysis, and the study of other aquatic microfossils that can provide independent evidence of limnologic change, it is possible to obtain some of this greatly needed specific information, especially if this approach is coupled with more detailed distributional and experimental studies.

Even if a detailed paleolimnologic reconstruction of the lakes investigated is not currently possible, the changes in diatom stratigraphy coincident with independent stratigraphic indicators of modern settlement near the lakes are in some cases so marked that they clearly indicate major lacustrine changes. These changes are briefly summarized below in the context of the lakes and their settlement history.

The taxonomy of the major species of these stratigraphic profiles is based principally on the works of Hustedt (1930), Patrick and Reimer (1966), and Cleve-Euler (1951-1955). Currently, however, there is some disagreement about the proper names for some of the minute members of the genus *Stephanodiscus*. Type material has not always been available for comparative purposes, and these taxonomic problems may take some time to resolve. Until then, I have chosen to follow the species concepts of E. F. Stoermer and his associates at the University of Michigan as applied to the diatom flora of Lake Michigan and Lake Ontario. This seems geographically logical, and it is hoped that the consistent terminology will facilitate communication among algal ecologists and paleolimnologists. Plates 1 to 6 illustrate the principal *Stephanodiscus* species found in this study. In several cases the *Stephanodiscus* species concepts used here differ from those reported earlier (for example, Bradbury and Megard, 1972; Bradbury and Waddington, 1973; Bright, 1968; Stark, 1971). Diagrams taken from these works have been changed accordingly after restudy of the prepared material. Nevertheless, the original epithets used in the cited literature are indicated in the figure captions.

Shagawa Lake

Shagawa Lake occupies an elongate glacial basin in bed rock in northeastern Minnesota (Fig. 1). It is 6 km long (east-west) and averages about 6 m deep, with maximum depths of 12 m. The lake lies immediately north of a once-rich deposit of iron ore of the Vermilion iron range. With the discovery of these deposits of hematite about 1886, the town of Ely rapidly developed. Land was cleared for mining development and home sites, and soon mine wastes, drainage from disturbed land, and finally domestic sewage entered the lake (Bradbury and Waddington, 1973).

The stratigraphic record of these events in a 160-cm core of profundal sediment consists of an abrupt rise of hematite silt at a depth of 34 cm that coincides with strong increases in the influx of pollen and fungal spores (Figs. 5, 6). This increase apparently results from the redeposition of pollen and fungal spores trapped in the litter on the forest floor when it was eroded following land clearance. The *Ambrosia* rise precedes these changes by about 5 cm and is presumed to have resulted from land clearance and lumbering far south of Shagawa Lake a decade or more earlier.

Before this time the pollen and sediment records are remarkably constant, and there is no indication of any unusual changes in the vegetation and landscape surrounding the lake back to about 15 A.D. (C^{14} date, 1935 ± 125 B.P.: I-6329) at the bottom of the core.

The presettlement limnology of Shagawa Lake is shown by the stratigraphic profiles of cladocerans (Fig. 6) and of diatoms (Fig. 7). The prevalence of the cladoceran *Chydorus sphaericus* and of the diatoms *Melosira ambigua, M. italica,* and *Fragilaria capucina* below 34 cm suggests that the lake was somewhat eutrophic in its natural state. Perhaps the nutrients that maintained the lake in this condition came from the leaching of patches of Pleistocene lake sediment that marks much of Shagawa Lake's drainage basin. The sharp relative increases in *F. capucina* at 40 and 100 cm (Fig. 7) indicate that the presettlement limnology was not entirely constant. These peaks may mean that the lake was more productive at those times or that there was a greater input of *F. capucina* from the littoral environment where it commonly lives to the profundal sediment.

Human disturbance of the landscape in the drainage basin of Shagawa Lake was accompanied by an increased flow of nutrients, leading to the greater dominance of *Melosira ambigua*. This was soon followed by a dominance of *Fragilaria crotonensis* and subsequently of *Stephanodiscus hantzschii* and *S. minutus* as domestic and municipal waste disposal into the lake increased (Bradbury and

Waddington, 1973). The maximum of *Stephanodiscus* spp. coincides with the rise of phosphorus (Fig. 6), presumably because of the introduction of phosphate detergent in about 1948 (Bradbury and Waddington, 1973).

The reduced dominance of the summer diatom, *Fragilaria crotonensis*, could be explained by massive blooms of blue-green algae that more efficiently reproduce

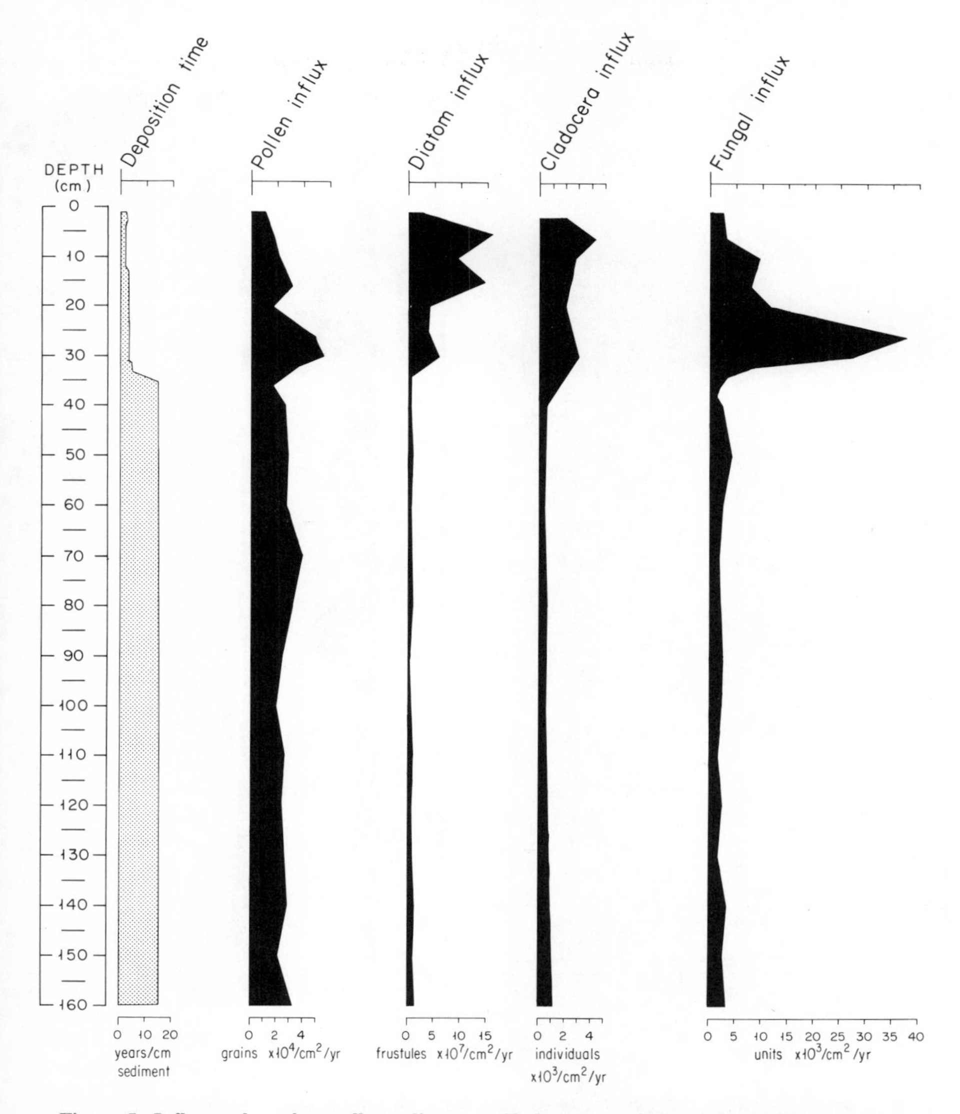

Figure 5. Influx values for pollen, diatoms, Cladocera, and fungi from Shagawa Lake, Minnesota (Bradbury and Waddington, 1973).

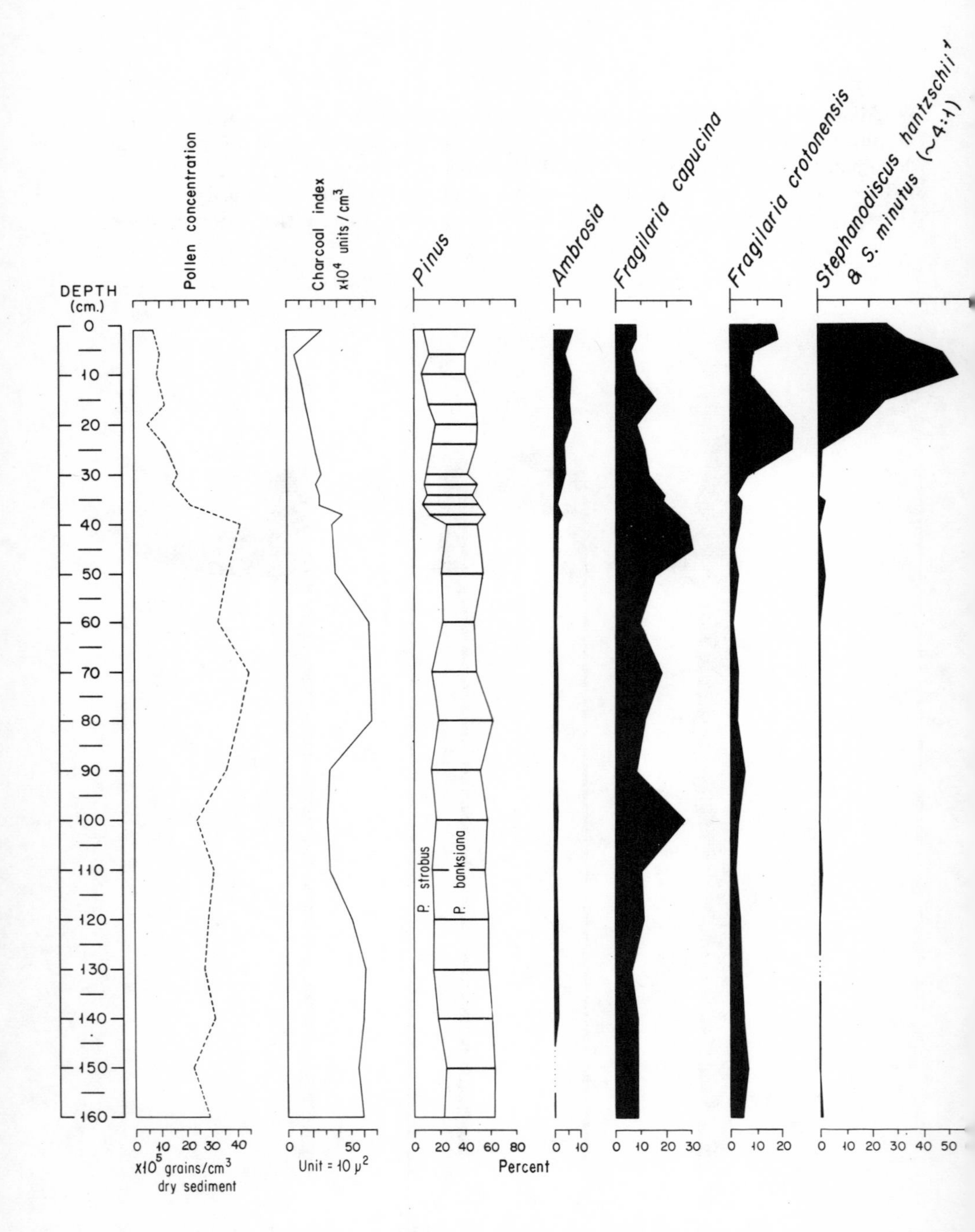

Figure 6. Selected stratigraphic profiles from Shagawa Lake, Minnesota. Modified from Bradbury and Waddington (1973).

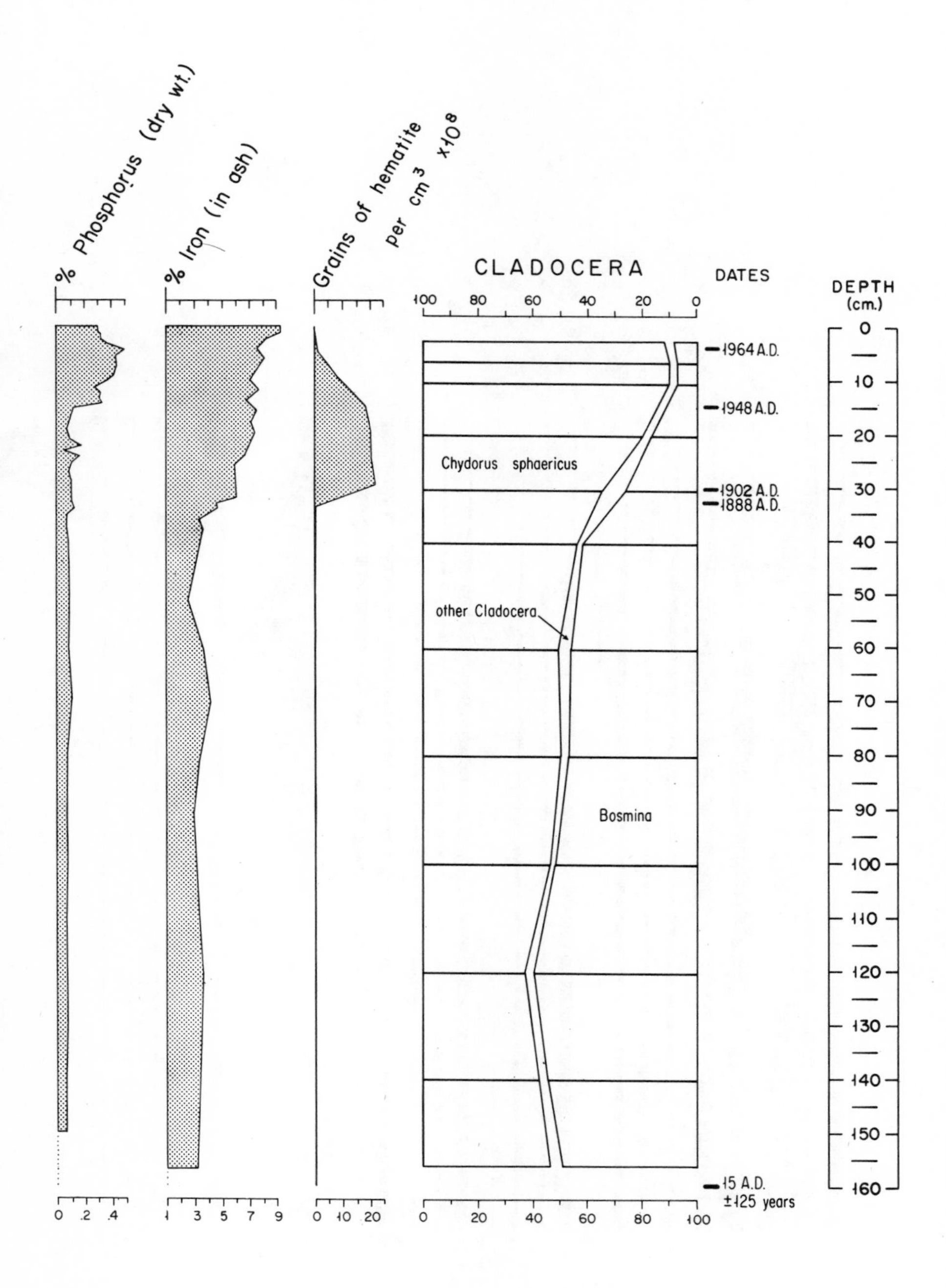

---

[1] Originally reported as *Stephanodiscus minutus.*

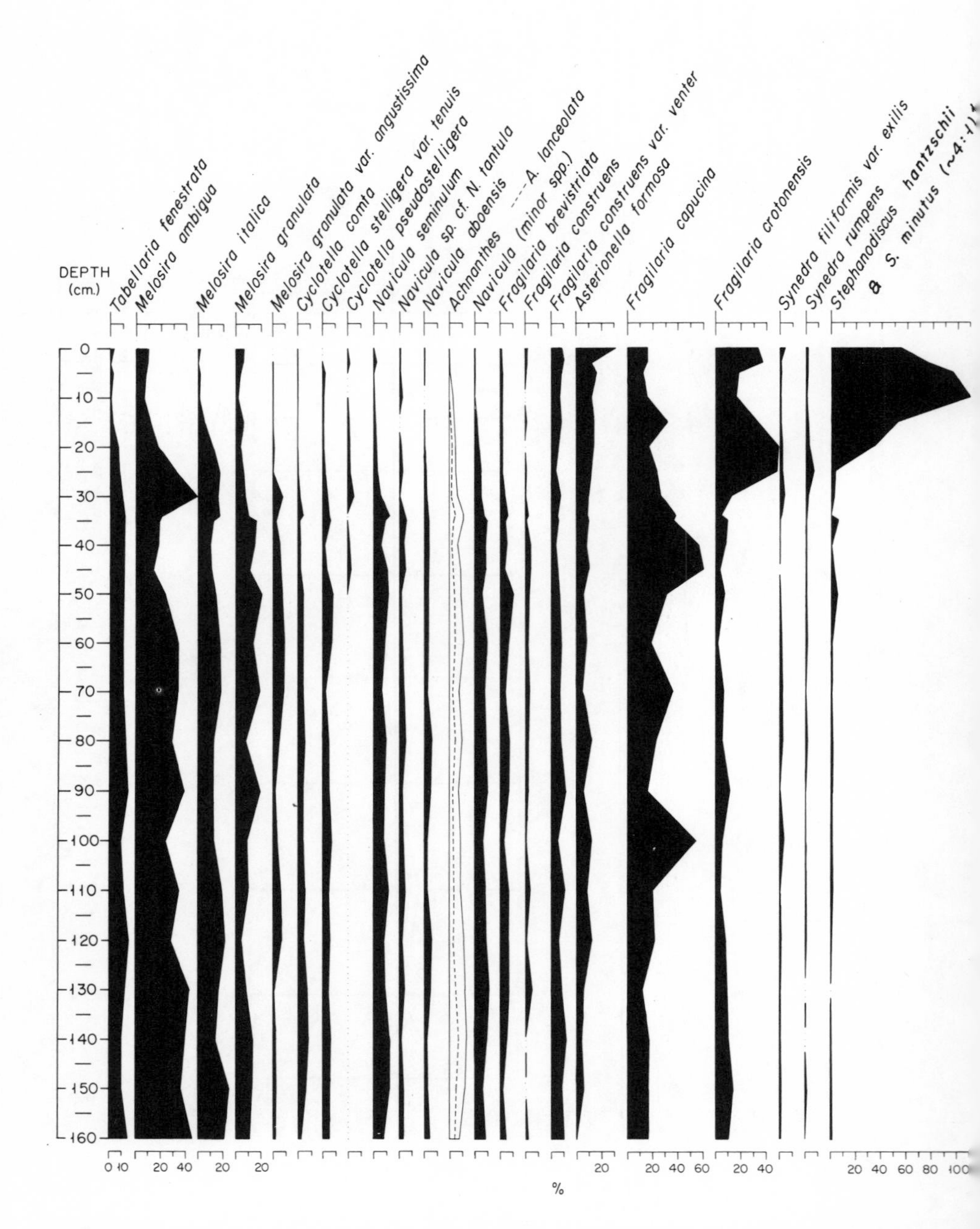

Figure 7. Diatom and sediment stratigraphy of Shagawa Lake, Minnesota. Modified from Bradbury and Waddington (1973).

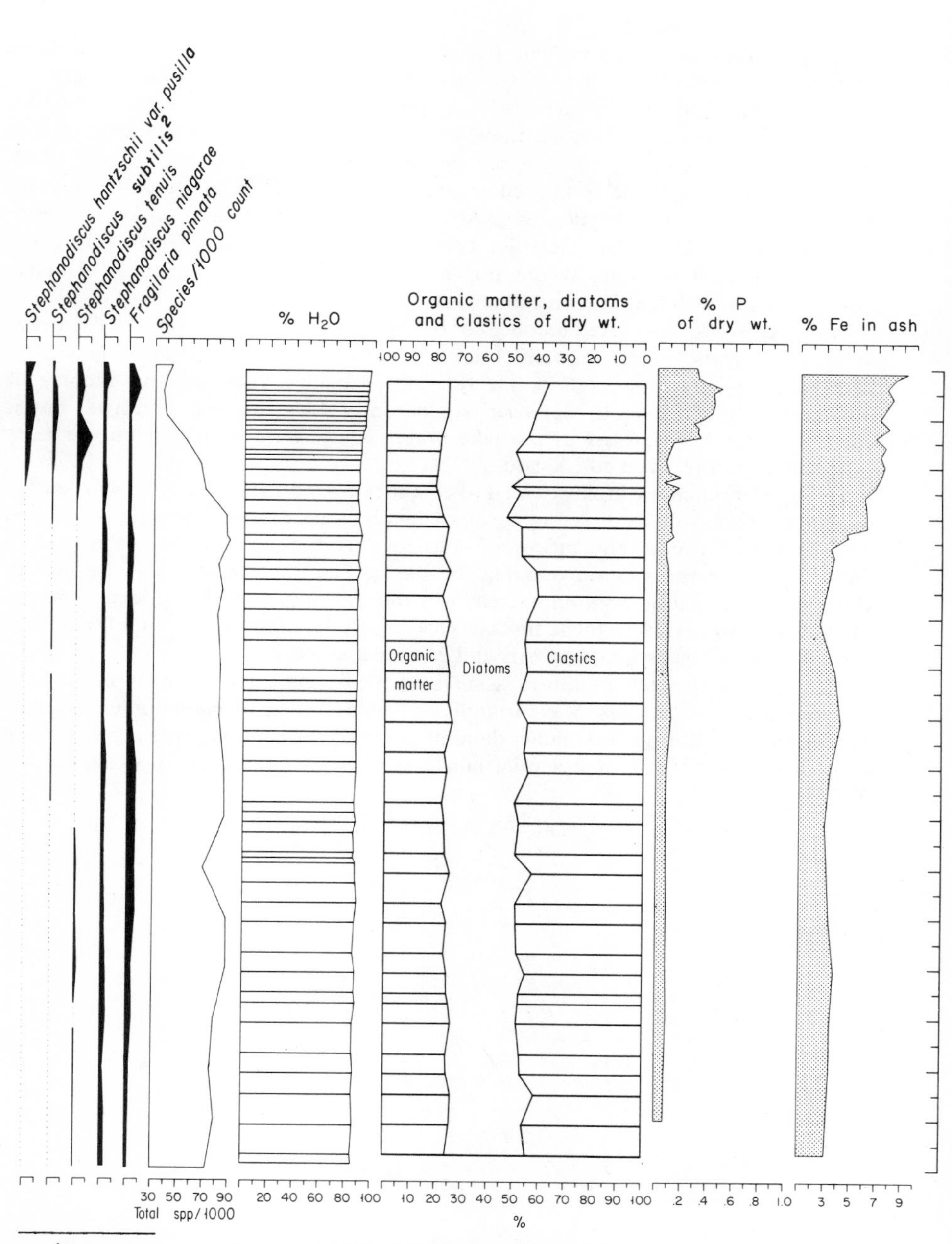

[1] Originally reported as *Stephanodiscus minutus.*

[2] Originally reported as *Stephanodiscus hantzschii.*

when phosphate is abundant. The blue-green algae not only compete with these diatoms for nutrients, but they also float and shade out the diatoms that are too heavy to stay near the surface. Both *Stephanodiscus hantzschii* and *S. minutus* avoid this competition because they tend to bloom in the early spring or late winter, sometimes even under the ice (Megard, 1969; Stoermer and others, 1974). Although more evidence is needed to accurately define the seasonality of these blooms, Megard's observation suggests that *S. hantzschii* prefers, or tolerates, low light levels. This characteristic, combined with its small (10-μm diam.) and easily suspended frustule, would make *S. hantzschii* and *S. minutus* effective competitors in water turbid with blue-green algae at any season.

The evidence for increased numbers of blue-green algae comes from the *Chydorus sphaericus* profile. This normally littoral benthic animal can become nominally planktonic by attaching itself to the floating filaments when massive blooms of blue-green algae occur. *C. sphaericus* apparently does not eat the algae, but it can feed on other detritus in the lake water while it rides around on the rafts of filaments (Bradbury and Megard, 1972).

The diatom changes in Shagawa Lake that occurred after the area was settled appear to relate to two different types of lake pollution. The first was associated with increased runoff and inflow of nutrients from natural sources that before were tied up in the soil surrounding the basin. The phytoplankton's response to this type of pollution was an increase in the eurytopic species already present in the lake, with a shift to those that are particularly favored by nutrient enrichment, for example, *Fragilaria crotonensis* and *Melosira ambigua*.

Massive phosphorus pollution associated with municipal wastes produces a different response. In this case eurytopic, nutrient-loving, spring-blooming diatoms are favored, but during the summer diatoms are replaced by blue-green algae because of competition or lack of essential nutrients such as silica (for an example, see Kilham, 1971).

Elk Lake

Elk Lake is a roughly oval lake about 1.6 km long (northwest-southeast) that stands in the headwaters of the Mississippi River in northern Minnesota (Fig. 1). The 30-m deep basin is of glacial origin, perhaps part of a once-ice-filled tunnel valley that was covered by drift during late Pleistocene time (Wright and Rhue, 1965).

The littoral sediments are silty to sandy biopelic marls, and the profundal sediments below water depths of approximately 20 m are laminated calcareous biopels (biopel ≅ gyttja, Bradbury and Waddington, 1973). The lamination results from light-colored laminae rich in calcium carbonate alternating in a regular pattern with possibly iron-rich gelatinous biopelic laminae. It suggests seasonal deposition in which each couplet of laminae represents an annual deposit or varve. The limnologic and sedimentologic mechanisms that produce the Elk Lake laminae are not specifically known, although similar sediments have been described from Ontario by Tippett (1964).

For the past 50 years, there has been no serious pollution of Elk Lake or damage to its surrounding landscape and forests. However, in the early 20th century, logging operations removed much of the forest cover near the lake. Beginning in 1901 dams were built and broken on Lake Itasca, downstream but at the same elevation as Elk Lake, to float logs down the Mississippi River. This activity backed water up Chambers Creek, which normally drains Elk Lake, and so raised and lowered the level of Elk Lake a metre or so from time to time. Such level changes probably occurred in the past, however, caused by beavers or drought, and may not have profoundly affected the limnology of Elk Lake. By 1917 logging reached the drainage basin of Elk Lake. Pine was selectively cut and hauled in the winter and stacked on the lake ice to await spring break-up and transportation downstream during the ice-free seasons. At that time a small dam was built at the outlet of Elk Lake (Chambers Creek) to provide water to float the logs downstream. During this time runoff over the partly denuded drainage basin and erosion of lake-margin soil by higher water levels may have provided an increased inflow of nutrients that enriched the lake to some extent (Stark, 1971). It is also likely that the nutrients and other organic compounds leached from logs stored in the lake (Schaumburg, 1973) would affect summer algal populations.

The stratigraphic record of logging is seen in a short core of profundal sediments as an increase in the sedimentation rate (Stark, 1971). Below 53 cm the deposition rate of regularly laminated sediment is about 3 mm/yr, whereas above this level the laminations are poorly developed and the deposition rate suddenly increases

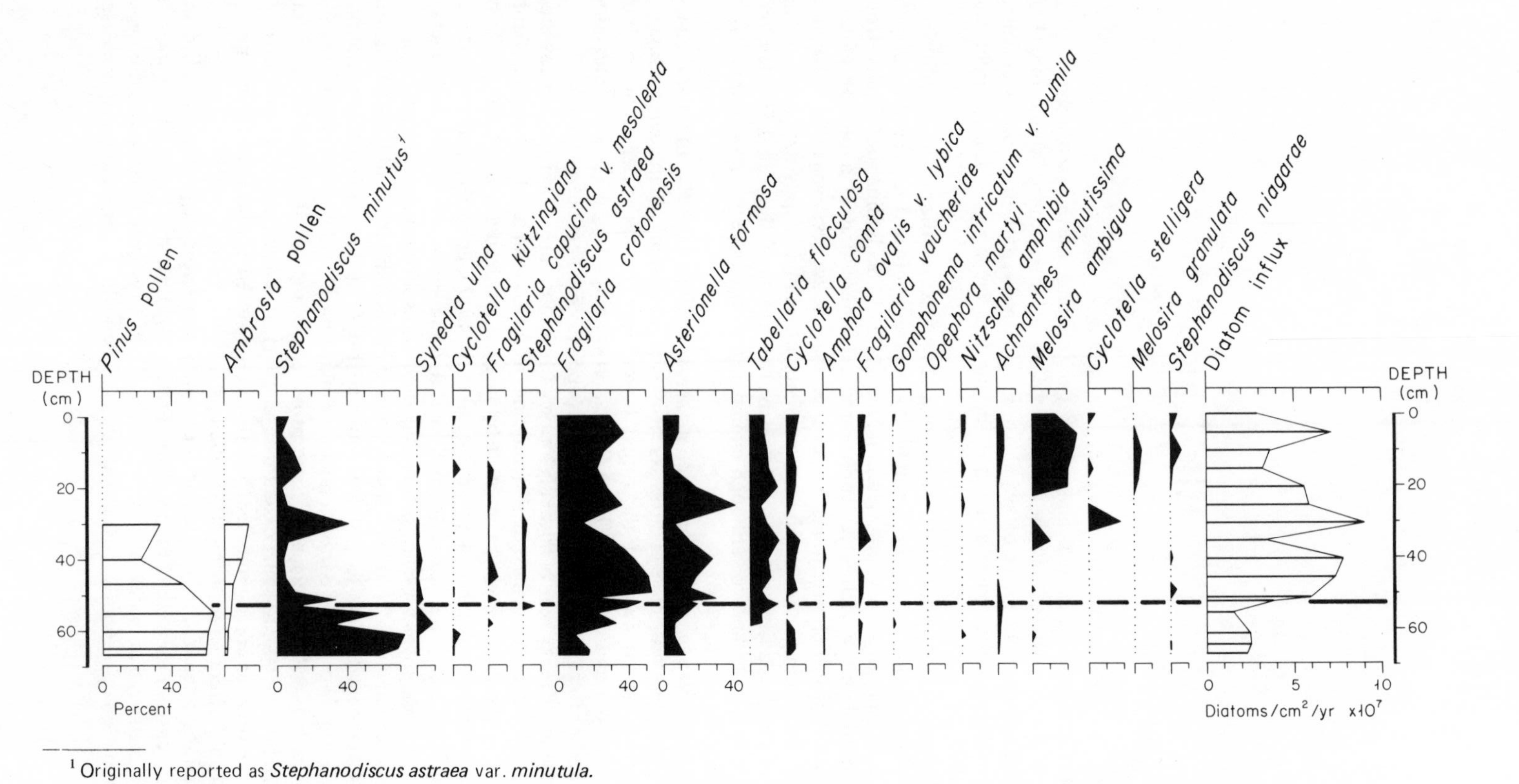

[1] Originally reported as *Stephanodiscus astraea* var. *minutula*.

Figure 8. Diatom stratigraphy from Elk Lake, Minnesota. Modified from Stark (1971).

to 10 mm/yr. This level is also marked by a drop in the percentage of *Pinus* pollen, which is presumed to result from logging in the region. About 3.5 cm below the transition to thick laminae, *Ambrosia* pollen begins its rise, marking settlement in northern Minnesota about 1890 (Fig. 8). Stark (1971) proposed that the increased productivity of the lake during the logging operation caused the higher deposition rate. In the top 14 cm of this core, the laminae become more closely spaced and more regular, suggesting decreased productivity following forest regeneration after logging.

The limnologic response of Elk Lake to these and other changes has been studied by means of numerous biostratigraphic analyses of several cores from shallow- and deep-water parts of the basin. Stark (1971) analyzed pollen, seeds, ostracods, chironomids, mollusks, and diatoms to determine the paleolimnology of Elk Lake. Diatom studies were limited to the profundal sediment.

The most obvious changes in the diatom stratigraphy (Fig. 8) occur through the transition from fine to coarsely laminated sediment. Below 53 cm *Stephanodiscus minutus* is the dominant diatom, but it is rapidly replaced by *Fragilaria crotonensis* above this level, although *S. minutus* becomes common once again at the 30-cm level for a brief period. Above 30 cm *F. crotonensis* remains the dominant diatom, joined by *Melosira ambigua*. Species of *Cyclotella* (*C. comta*, *C. stelligera*, and *C. kutzingiana*) and *Asterionella formosa* have a rather erratic stratigraphic distribution, but *Tabellaria flocculosa* (= Stark's *T. fenestrata*) persists without much change from 53 cm to the surface.

The stratigraphy is difficult to interpret because the species undergo rapid fluctuations from high to low levels, and by and large they are ecologically eurytopic and thus do not yield easily to simple ecologic interpretation. The presence of *Stephanodiscus minutus* and *Fragilaria crotonensis* (53 to 67 cm) suggests that Elk Lake was reasonably productive just before the presumed logging disturbance, and the few analyses (at 1-m intervals) from a long (13 m) core (Fig. 9) show that these species were common for the past 7,000 yr. On the other hand, the presence of *Cyclotella comta*, *Tabellaria flocculosa*, and *Asterionella formosa* certainly indicates that Elk Lake was far from being polluted in the sense of Shagawa Lake or other lakes that will be discussed. These species, while considered eurytopic in Lake Michigan (Stoermer and Yang, 1970), are characteristic of the less productive lakes of northern Minnesota (Fig. 3).

Despite the difficulty of interpreting the Elk Lake diatom stratigraphy in detail, it is possible to construct a reasonable hypothesis that accounts for the major changes. In Elk Lake, like Shagawa Lake, *Stephanodiscus minutus* probably reaches its highest populations in late winter and early spring, sometimes blooming under the lake ice. Later in the spring and in early summer, several other species become prominent, such as *Asterionella formosa* and *Tabellaria flocculosa*, and finally in middle to late summer, *Fragilaria crotonensis* blooms. Although the seasonal succession of planktonic diatoms is subject to considerable variation from lake to lake and year to year, this sequence is probably correct in a general way for eutrophic-mesotrophic, north temperate lakes in continental climates. Hutchinson's (1967) summary of data on the seasonal development of phytoplankton gives this impression, even though much of his discussion concentrates on variations in the pattern.

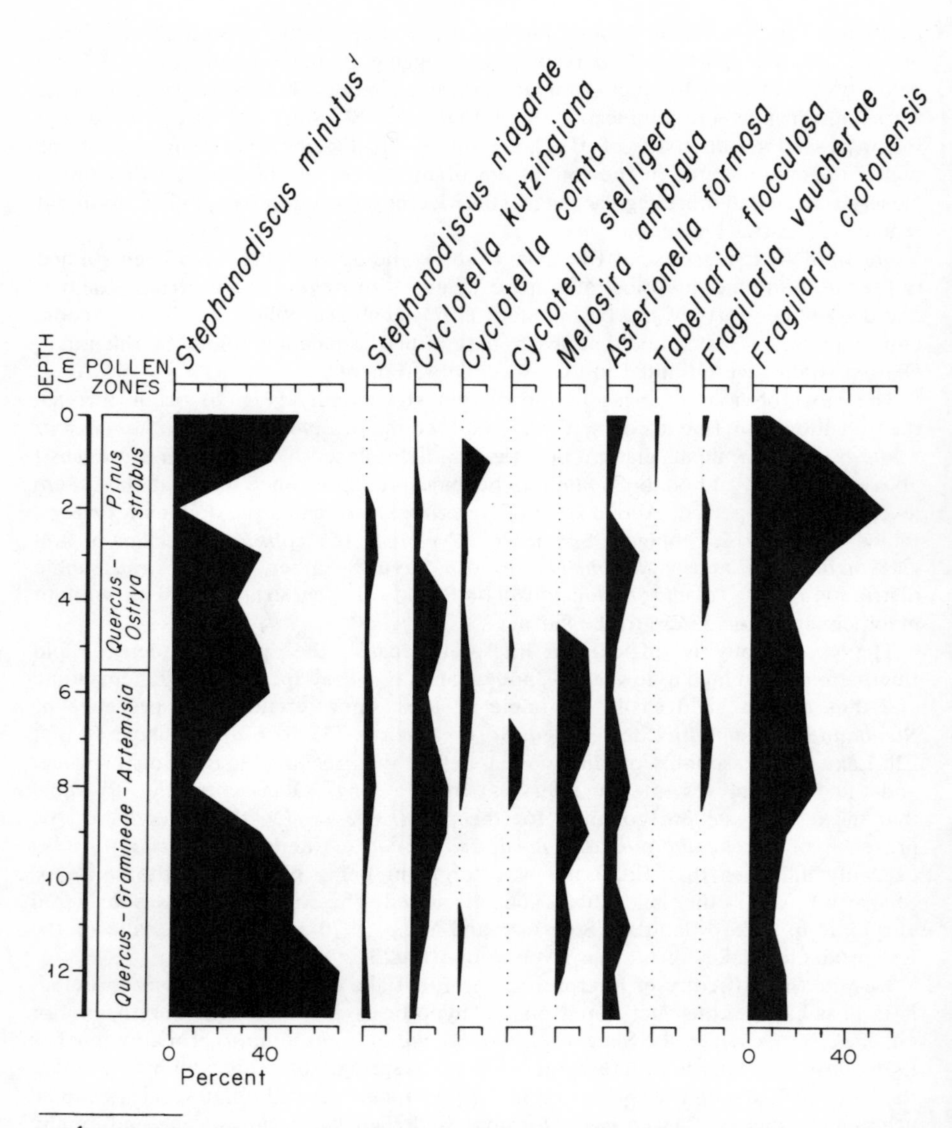

[1] Originally reported as *Stephanodiscus astraea* var. *minutula.*

Figure 9. Diatom stratigraphy of the long core (69-6), Elk Lake, Minnesota. Modified from Stark (1971).

bance levels are characterized by littoral diatoms. In Sallie A this zone has more planktonic diatoms and a greater variety of littoral forms than it does in Sallie P, where *Amphora ovalis* var. *pediculus* is the constant but waning dominant, along with *Cocconeis* cf. *diminuta*. At the *Ambrosia* rise, the littoral flora is replaced by increasing numbers of planktonic diatoms, notably *Melosira ambigua, M. granulata,* and *Cyclotella comta,* and these in turn are replaced or joined by a massive development of *Stephanodiscus hantzschii,* which continues to dominate the upper levels. Some littoral species persist throughout the record (for example, *Fragilaria pinnata*), and several increase after the *Ambrosia* rise, such as *F. capucina* var. *mesolepta, F. brevistriata, F. construens* var. *binodis,* and *F. vaucheriae*.

A general interpretation of these stratigraphic changes is consistent with the pattern of diatom development in other disturbed lakes; namely, a predisturbance diatom assemblage, usually with a relatively greater representation of littoral species, is replaced by planktonic forms that thrive in eutrophic water. Initially, increases in nutrient levels encourage both spring- and summer-blooming diatoms, but as enrichment continues, blue-green algal blooms in middle to late summer eliminate the summer diatom plankton, perhaps by competition for light. *Stephanodiscus hanztschii,* unaffected by this competition because it blooms in the late winter or early spring, persists and dominates the stratigraphic record. The chydorid *Chydorus sphaericus* can be taken as an indicator of blue-green algal blooms in lakes because its numbers increase when it can feed in the limnetic area of a lake by hitching rides on floating blue-green algal filaments or colonies (Bradbury and Megard, 1972). The profile for *C. sphaericus* in core Sallie P (Fig. 12) clearly correlates with *S. hantzschii* and provides the rationale for implicating blue-green algal competition as an important factor in determining the planktonic diatom assemblages.

Changes in the stratigraphic distribution of the littoral diatoms may not directly reflect increasing nutrient levels in Lake Sallie. The predisturbance dominant *Amphora ovalis* var. *pediculus* is primarily a benthic littoral diatom (Fig. 2) that sometimes attaches itself epiphytically to other large diatoms such as *Surirella* and *Nitzschia* species (Hustedt, 1930). Its small size and rounded dorsal contour suggest that it might tolerate high-energy environments, and in Denmark it is dominant in the surf zone of many eutrophic lakes (Jorgensen, 1948). *Cocconeis diminuta* occurs in similar habitats, but it is not as common.

The littoral species of *Fragilaria* are epiphytic or benthic. During cell division these diatoms form long ribbons of individuals that become loosely attached to convenient substrates. This mode of existence does not appear to be well adapted to shallow, open, and highly agitated aquatic environments, and it is not surprising to find them preferring quiet littoral environments where water currents and waves are slowed by growth of aquatic macrophytes. Their distribution in the deeper littoral areas of Danish lakes and ponds supports this generalization (Jorgensen, 1948).

Against this background, the transition from *Amphora ovalis* var. *pediculus* and *Cocconeis diminuta* in the predisturbance zone to *Fragilaria* species in the postdisturbance levels seems likely to be the result of a rise in water level. As lake levels rose, the shallow-water, high-energy habitats would have transgressed away from the site of deposition represented by the core and would have been replaced by deeper, quieter water environments in which the *Fragilaria* species could prosper. After disturbance, as nutrient levels increased in Lake Sallie, massive growths of macrophytic vegetation would provide additional habitats in quiet water for *Fragilaria* species. Such vegetation has plagued Lake Sallie for some time (Mann and McBride, 1972), and dense growths of aquatic vegetation presently grow on about 34 percent of the bottom area of the lake (Neel and others, 1973).

The correlation between these limnologic changes and European settlement in the region seems reasonably well established. Nevertheless, it is apparent that the transition from a littoral environment dominated by *Amphora ovalis* var. *pediculus* to one of increased planktonic diatoms and species of *Fragilaria* began considerably before the *Ambrosia* rise marks the beginning of intensive European settlement.

Will (1946) reported that a severe drought in northwestern Minnesota ended in 1850, and the low lake levels began to rise. The predisturbance record of *Amphora ovalis* var. *pediculus* may reflect this rising trend in lake level. Below 60 cm in the Sallie P core (length = 75 cm), diatoms are absent or poorly preserved. If the bottom of the core represents the 1850 drought, as suggested by H. H. Birks, M. C. Whiteside, D. M. Stark, and R. C. Bright (in prep.), then the low lake levels may explain the general scarcity of well-preserved diatoms in this zone. When lake levels are low, sediments are aerated and agitated, and diatoms may preserve poorly, perhaps because of mechanical breakage and (or) ingestion by benthic scavengers. In addition, as the lake lowered, shelf sediments in the littoral area—poor in diatoms according to the above reasoning—would be reworked and transported over the shelf edge to the core-site location, diluting the normal diatom sedimentation. An anomalously high sedimentation rate is suggested for this part of the Sallie P core. Considering the *Ambrosia* rise to represent 1870 and the base to represent 1850 (the end of the drought), the rate becomes about 1.5 cm/yr. Above the *Ambrosia* rise, the average sedimentation rate is only 0.45 cm/yr, but this represents mostly limnetic deposition.

In the St. Clair Lake core, *Ambrosia* pollen begins to increase above background levels at a depth of 45 cm (Fig. 13) and probably represents agricultural disturbance in the early 1870s, when the first grain crops were sown. The large increase in *Ambrosia* pollen at 33 cm may represent the drainage of St. Clair Lake in 1915 and the colonization of the exposed lakeshore sediments by *Ambrosia* and other herbs.

The diatom stratigraphy of St. Clair Lake (Fig. 13) is basically similar to that of Lake Sallie. Predisturbance littoral diatoms change through the compound *Ambrosia* rise to postdisturbance eutrophic planktonic diatoms, first *Stephanodiscus hantzschii* and later *Melosira granulata* var. *angustissima*. The predisturbance littoral flora contains many species of *Navicula*, *Cymbella*, and other genera that collectively dominate along with *Fragilaria brevistriata*. Motile benthic diatoms such as the *Navicula* species could exist throughout the lake in nonvegetated or sparsely vegetated areas. Pollution of such a lake first by agricultural drainage and later by municipal wastes might eliminate the open areas of the lake bottom and its benthic flora by dense growths of submerged aquatics. Additionally, as the input of organic matter increased and the sediments became more reducing, less tolerant benthic species would not survive.

Planktonic eutrophic species, *Melosira granulata* var. *angustissima* and *Stephano-*

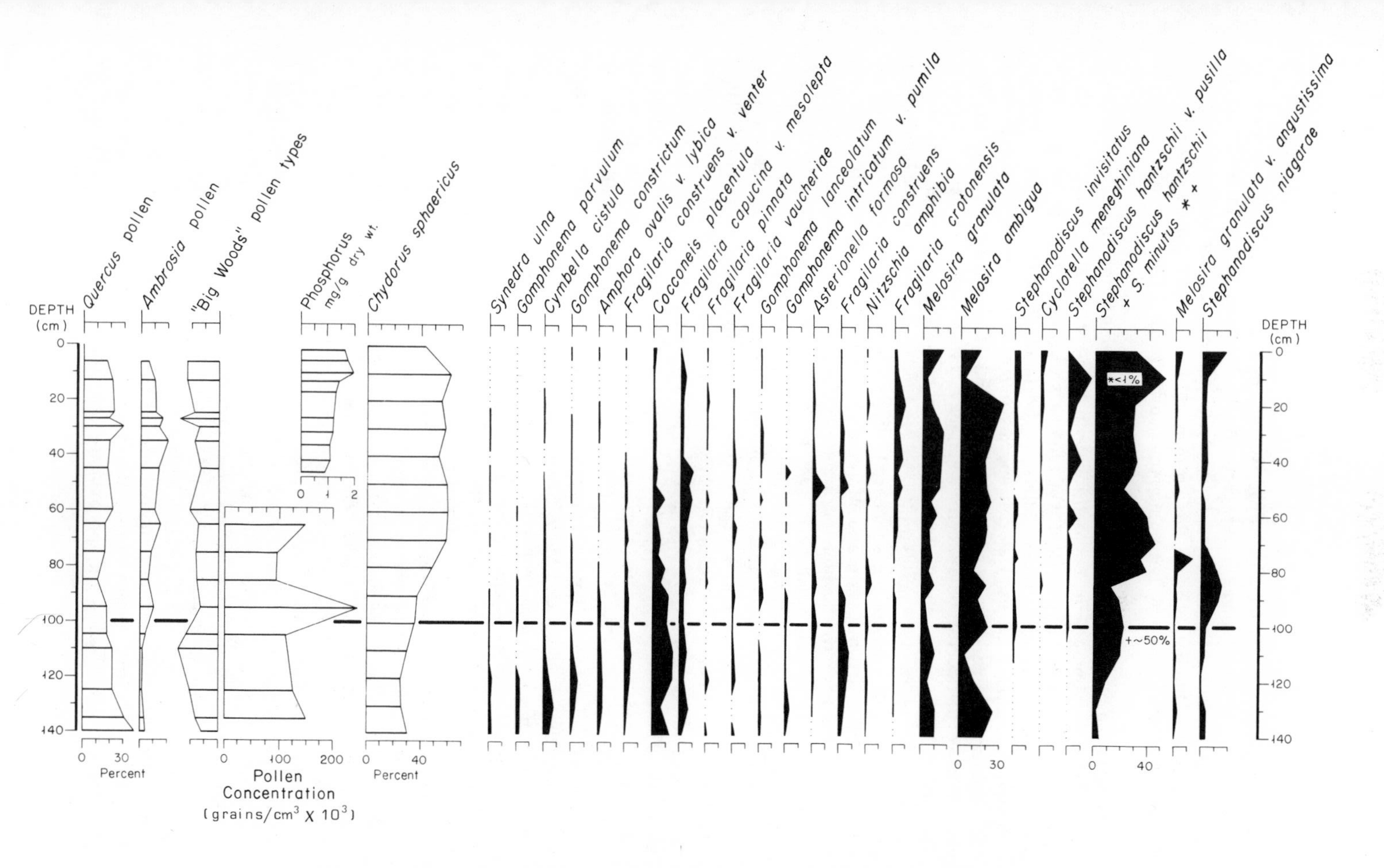

Figure 14. Diatom stratigraphy of Tanager Lake, Minnesota. Analyst: K. K. Baker, 1971.

The profile for the cladoceran *Chydorus sphaericus* shows a significant increase above 90 cm in the Tanager Lake core. As in Shagawa Lake, the increase in *C. sphaericus* indirectly indicates an increase in the number and intensity of blue-green algal blooms because this normally littoral animal can expand into the limnetic area of the lake by becoming attached to floating filaments of blue-green algae.

According to a model established earlier, summer-blooming planktonic diatoms should be disadvantaged in competition with blue-green algae because of their greater weight and inability to float. Only in cases where lake morphometry insures active summer turbulence that extends to the lake bottom can heavy summer planktonic diatoms such as *Melosira* species maintain reasonable populations in the plankton. The persistence of *M. granulata* and *M. ambigua* in Tanager Lake, along with evidence for blue-green algae, may reflect these special conditions. Tanager Lake is small (about 400 m in diameter) and comparatively shallow (6 m), and although the lake may stratify during the day, it is probably easily mixed on windy days throughout the summer. Under such conditions, diatoms of many kinds, including *M. granulata* and *M. ambigua,* would be recirculated from the nearby littoral zone to the limnetic area. In addition, Long Lake Creek, which flows into this relatively small basin, may be capable of introducing such diatoms. Turbulence caused by boating activity may also represent an important factor.

Stephanodiscus hanztschii reaches its maximum concentration at a depth of 10 cm, coincident with the major peak in the abbreviated phosphorus profile (Fig. 14). This same relationship occurs in Shagawa Lake (Fig. 6) and probably reflects the same conditions—massive blue-green algal blooms that displace all other diatom plankton. The peak in *S. hantzschii* (+ <1 percent *S. minutus*) is preceded by decreases in *Melosira granulata* and *M. ambigua.* This may result simply from an exceptionally large production of *S. hantzschii* that proportionally produces decreases in other species, or following the hypothesis outlined above, the decrease in summer diatom plankton (*Melosira* species) may reflect decreased turbulence. This may happen by meteorologic circumstance—comparatively calm summers, for example—or it may result from greater lake stability caused by blue-green algae. As blue-green algae increase, the water becomes denser optically and hence better able to absorb heat from summer insolation. This effect has been achieved experimentally by coloring water of a Nebraska farm pond with dyes (Buglewicz, 1972). Under such conditions, a strong stratification develops, with warm, light water at the surface, and it becomes more difficult to mix even small, shallow lakes on windy days. As the lake becomes more stable, the diatom phytoplankton that depends on turbulence for suspension finally settles out.

The upper 10 cm of the Tanager Lake record contain evidence for decreased pollution, as indicated by the declining numbers of *Chydorus sphaericus, Stephanodiscus hantzschii,* and the drop in the phosphorus profile. *Melosira* species again join the summer phytoplankton. Sewage-treatment facilities were finally completed for the Long Lake community in 1964 (Minnesota Pollution Control Agency, 1972). Before that time individual households used septic tanks and drain fields, and most of the effluent ultimately entered Long Lake Creek and then Tanager Lake. The level of the first *Ambrosia* rise (100 cm $\cong$ 1860) and the pollen concentration peak (95 cm $\cong$1870) allow an approximate calculation of the deposition time for the postdisturbance Tanager Lake sediments: 1 yr/cm. At this rate a depth of 10 cm would represent the early 1960s, and the stratigraphic evidence suggesting reduced phosphorus levels and blue-green algal blooms would correspond to the construction of the sewage treatment plant from 1963 to 1964.

BROWNS BAY

Although Browns Bay lies adjacent to Tanager Lake and probably had a similar settlement history, its sedimentary record differs in some important aspects. The Browns Bay core came from the steep northeast littoral slope of the bay beneath about 14 m of water (Fig. 1). In this type of lacustrine environment, sediment deposition alternates with periods of subaqueous erosion, and the net rate of sediment accumulation is low, because fine organic detritus is winnowed to greater depths. The *Ambrosia* rise occurs at 14 cm (Fig. 15) and probably represents the early 1860s as in Tanager Lake. In the 110 years that followed, a biopelic sandy silt with some gravel accumulated at an average rate of 1.3 mm/yr. Coarse clastic sediments continue below the *Ambrosia* rise, but below 40 cm, sediments are more organic and suggest more stable lake-bottom conditions and sedimentation patterns. In contrast to Tanager Lake, the curve for pollen concentration does not show a significant increase at or after the *Ambrosia* rise, as might be expected if pollen from forest litter were being washed into the bay as erosion increased following clearing. Most likely this results because small sediment particles like pollen grains are transported down the littoral slope during periods of turnover, as Davis (1973) suggested. The drop in the pollen concentration curve above 14 cm may represent an increase in the rate of autochthonous sediment deposition resulting from cultural eutrophication.

The diatom stratigraphy of Browns Bay reflects this eutrophication in much the same way it has in other disturbed lakes. The predisturbance diatom flora (below 14 cm) consists mostly of littoral species such as *Fragilaria construens* var. *venter*, *Amphora ovalis* var. *lybica*, *A. ovalis* var. *pediculus*, and *Synedra ulna*. Associated with this flora are planktonic diatoms characteristic of nonenriched Minnesota lakes—*Tabellaria flocculosa*, *Cyclotella kutzingiana*, and *C. comta*—and eurytopic planktonic species like *Melosira granulata* and *M. ambigua*. After enrichment, probably from cultural wastes entering through Tanager Lake and from nearby lakeshore dwellings, the littoral species decline, and eurytopic, enrichment-favored planktonic species (*Stephanodiscus minutus*, *S. hantzschii*, and *Melosira ambigua*) replace them. *Fragilaria capucina* and *F. capucina* var. *mesolepta*, which are littoral or meroplanktonic diatoms of enriched lakes, also increase at this time.

Significant changes in the diatom stratigraphy below the *Ambrosia* rise, such as the partial replacement of *Fragilaria construens* var. *venter* by *Melosira ambigua* and *M. granulata* between 55 and 30 cm, are difficult to relate to presettlement events. A pollen diagram for Browns Bay (Fig. 16) shows a considerable increase in the pollen of *Ostrya* or *Carpinus* and *Ulmus* between 35 cm and the *Ambrosia* rise (14 cm); this may represent the return of the Big Woods vegetation to this area about 400 yr ago (Waddington, 1969). The high oak pollen values suggested by the few analyses below 20 cm implies the existence of an oak savannah surrounding

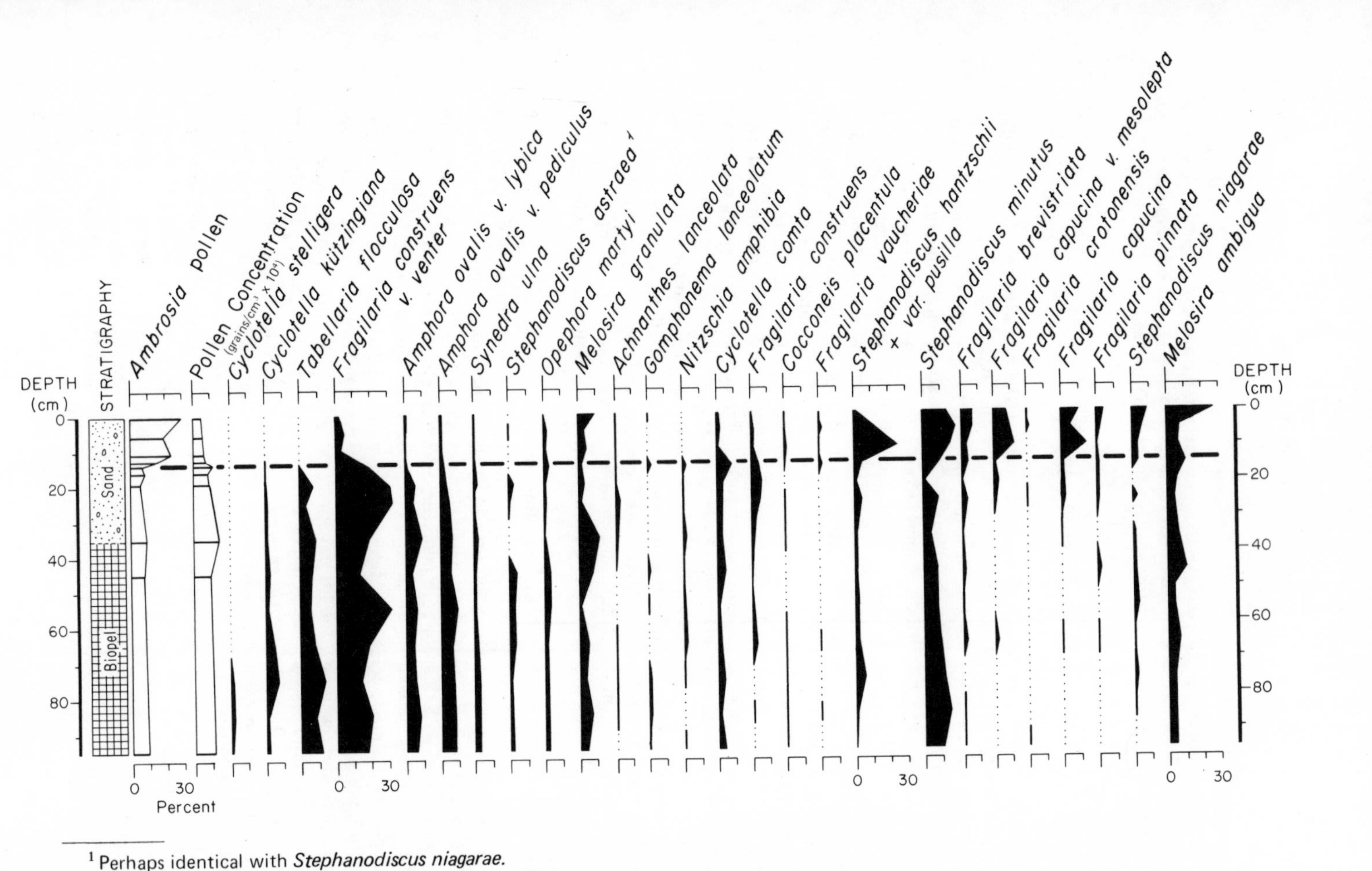

[1] Perhaps identical with *Stephanodiscus niagarae.*

Figure 15. Diatom stratigraphy of Browns Bay, Lake Minnetonka, Minnesota. Analyst: N. P. Sather, 1971.

Conclusions

The recent stratigraphy of diatom assemblages in several Minnesota lakes shows marked changes that correlate with independent stratigraphic indications of European-American settlement in the lake's drainage basin and around its shores. By ecologic comparison with modern diatom floras, the fossil diatom assemblages indicate that the lakes became more productive. The correlation of these changes with settlement horizons suggests that lake enrichment resulted from increased nutrients released by soil erosion following land clearance, and frequently later from nutrients supplied by municipal and private waste disposal into the lakes.

Our ability to interpret the details of lake enrichment, including the specific kinds and amounts of pollutants, depends upon our understanding of the nutritional, seasonal, and distributional ecology of the diatom species and associations and upon the accuracy of correlation with historic settlement events. In many cases this knowledge is inadequate, and detailed paleolimnologic interpretations are not possible.

Ecologic information on diatom associations will be gathered in both field and laboratory investigations, and the stratigraphic record of diatom assemblages will provide us with historical insights about the persistence and variability of freshwater diatom floras and the general limnologic implications that will be important in the autecologic assessment of a given species. Stratigraphic refinement can be accomplished by combining a thorough review of the history of local settlement patterns and events with detailed stratigraphic analysis of pollen, mineral assemblages, and sediment chemistry that directly relates to local environmental changes.

This stratigraphic study of several lacustrine environments in Minnesota and South Dakota suggests that the reaction of diatom floras to cultural disturbance varies according to the lake's limnology and the amount and timing of influxes of different nutrients. Because of this, generalizations about the specific changes that have taken or are taking place are unlikely to be of great value in assessing pollution problems in other lakes. Only by studying the transition of a lake from natural to disturbed conditions is it possible to understand or at least appreciate the impact on lakes of intensive settlement. Nevertheless, it is perhaps worthwhile, even at this early stage, to summarize the more obvious changes in the diatom flora as lakes become highly enriched. Table 4 compares a generalized natural lake with one that has been culturally enriched as in the highly eutrophic environments of this study. It is conceivable that similar and perhaps identical changes could occur naturally.

TABLE 4. SUMMARY OF GENERAL CHANGES IN THE DIATOM FLORAS OF ENRICHED LAKES

	Predisturbance assemblage	Postdisturbance assemblage
Diatom diversity	Tends to be higher both in regard to planktonic and littoral diatoms	Tends to be lower, with only a few species dominating
Planktonic	There is a more or less even representation of spring, summer, and fall species	With enrichment there is a shift to spring- and perhaps late fall-blooming species as other planktonic algae, particularly blue-green algae, prosper in middle summer
	Usually several eurytopic species are present	One or few eurytopic species predominate. *Stephanodiscus hantzschii* characterizes the most eutrophic lakes, while *S. minutus* is generally found in lakes that have received less enrichment
Benthic and epiphytic species	There is a greater diversity of types and greater numbers of individuals	The benthic flora is proportionally reduced after disturbance. Epiphytic diatoms may increase with increased aquatic macrophytes, but often they are swamped by massive blooms of planktonic diatoms
Diatom productivity	Varies according to trophic state of the lake	Increases with enrichment of limnetic environment
Total flora	Represents a distinctive adjustment to particular limnologic conditions	Highly enriched lakes generally have similar diatom floras. The species are not unique to polluted lakes

PLATE SECTION

PLATE 1

Light microscope photographs of *Stephanodiscus* species from Minnesota lake sediment

Figure 1. ***Stephanodiscus hantzschii* Grunow. Diameter = 10 μm. Shagawa Lake, 15 cm**
Figure 2. ***Stephanodiscus hantzschii* Grunow. Diameter = 10 μm. Tanager Lake, 10 cm**
Figure 3. ***Stephanodiscus minutus* Cleve and Möller. Diameter = 7.5 μm. Shagawa Lake, 15 cm**
Figure 4. ***Stephanodiscus minutus* Cleve and Möller. Diameter = 9 μm. Tanager Lake, 10 cm**
Figure 5. ***Stephanodiscus subtilis* Van Goor. Diameter = 10 μm. Tanager Lake, 10 cm**
Figure 6. ***Stephanodiscus subtilis* Van Goor. Diameter = 9 μm. Shagawa Lake, 15 cm**
Figure 7. ***Stephanodiscus tenuis* Hustedt. Diameter = 8 μm. Shagawa Lake, 15 cm**
Figure 8. ***Stephanodiscus tenuis* Hustedt?. Diameter = 9 μm. Tanager Lake, 10 cm**
Figure 9. ***Stephanodiscus dubius* Fricke. Diameter = 11 μm. Tanager Lake, 10 cm**
Figure 10. ***Stephanodiscus dubius* Fricke. Diameter = 6.5 μm. Shagawa Lake, 15 cm**
Figure 11. ***Stephanodiscus niagarae* Ehrenberg. Diameter = 34 μm. Tanager Lake, 10 cm**
Figure 12. ***Stephanodiscus niagarae* Ehrenberg. Magnification = 2,000×. Tanager Lake, 10 cm**

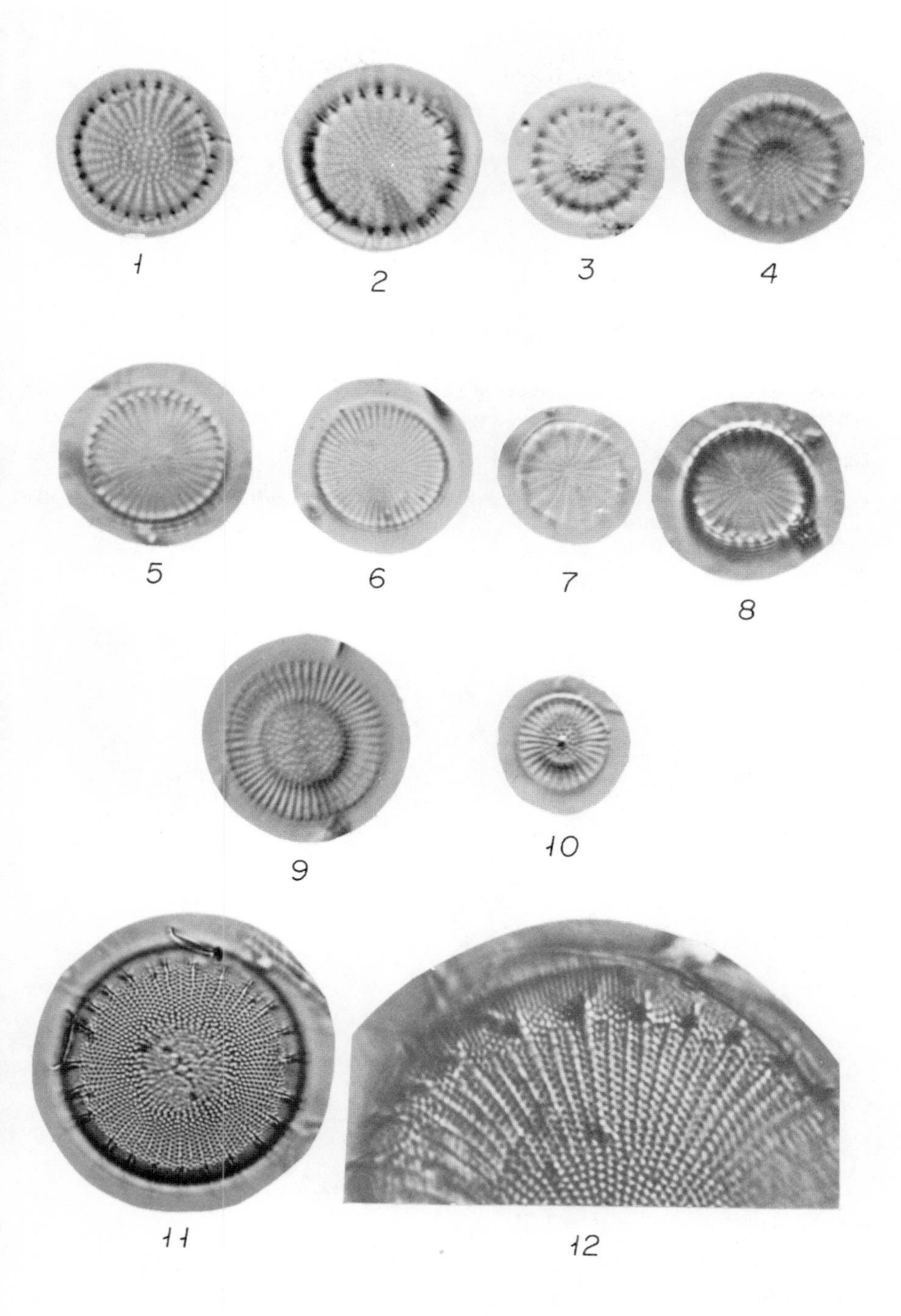
1
2
3
4
5
6
7
8
9
10
11
12

PLATE 2

Scanning electron microscope photographs of *Stephanodiscus* species from Minnesota lake water and sediment

Figure 1. ***Stephanodiscus tenuis*** **Hustedt (external view). Diameter = 9 μm. Shagawa Lake plankton**

Figure 2. ***Stephanodiscus niagarae*** **Ehrenberg (internal view). Diameter = 60 μm. Sallie Lake sediment, 1.5 to 4.5 cm**

1

2

PLATE 3

Scanning electron microscope photographs of *Stephanodiscus niagarae* Ehrenberg from sediment of Sallie Lake, Minnesota

Figure 1. ***Stephanodiscus niagarae*** **Ehrenberg (detail of punctae). Magnification = 10,000×. Sallie Lake sediment, 1.5 to 4.5 cm**

Figure 2. ***Stephanodiscus niagarae*** **Ehrenberg (external view). Diameter = 48 μm. Sallie Lake sediment, 1.5 to 4.5 cm**

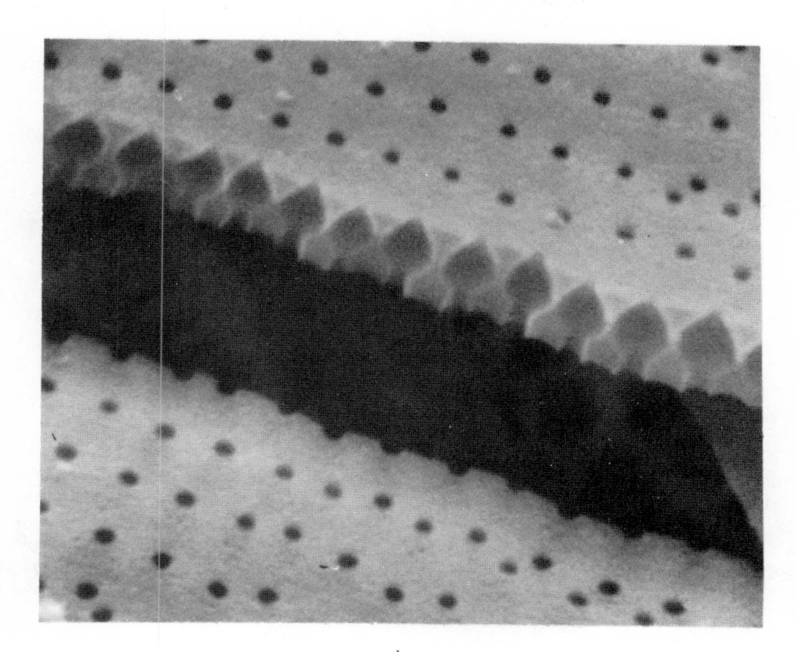

1

2

PLATE 4

Scanning electron microscope photographs of *Stephanodiscus* species from the plankton of Shagawa Lake, Minnesota

Figure 1. ***Stephanodiscus dubius*** **Fricke (internal view). Diameter = 6 μm. Shagawa Lake plankton**
Figure 2. ***Stephanodiscus dubius*** **Fricke (external view). Diameter = 6 μm. Shagawa Lake plankton**
Figure 3. ***Stephanodiscus subtilis*** **Van Goor (external view). Diameter = 8 μm. Shagawa Lake plankton**

References Cited

Bradbury, J. P., and Megard, R. O., 1972, A stratigraphic record of pollution in Shagawa Lake, northeastern Minnesota: Geol. Soc. America Bull., v. 83, p. 2639-2648.

Bradbury, J. P., and Waddington, J.C.B., 1973, The impact of European settlement on Shagawa Lake, northeastern Minnesota, U.S.A., *in* Birks, H.J.B., and West, R. G., eds., Quaternary plant ecology: Oxford, Blackwells, p. 289-307.

Bright, R. C., 1968, Surface-water chemistry of some Minnesota lakes, with preliminary notes on diatoms: Minnesota Univ. Limnol. Research Center Interim Rept. 3, 58 p.

Buglewicz, E. G., 1972, The impact of reduced light penetration on a eutrophic farm pond [M.S. thesis]: Lincoln, Univ. Nebraska, 99 p.

Cholnoky, B. J., 1968, Die Okologie der Diatomeen in Binnengewassern: Weinheur, J. Cramer, 699 p.

Cleve-Euler, Astrid, 1951-1955, Die Diatomeen von Schweden und Finland (Teil I-V): Kungl. Svenska. Vetenskapsakademiens Handlinger (Fjärde Serien); Teil I, Band 2, No 1 (1951), 163 p.; Teil II, Band 4, No 1 (1953), 158 p.; Teil III, Band 4, No 5 (1953), 255 p.; Teil IV, Band 5, No 4 (1955), 232 p.; Teil V, Band 3, No 3 (1952), 153 p.

Davis, M. B., 1970, Erosion rates and land use history in southern Michigan: Geol. Soc. America Abs. with Programs, v. 2, p. 533.

——1973, Redeposition of pollen grains in lake sediment: Limnology and Oceanography, v. 18, p. 44-52.

Evans, D., and Stockner, J. G., 1972, Attached algae on articifial and natural substrates in Lake Winnipeg, Manitoba: Fisheries Research Board Canada Jour., v. 29, p. 31-44.

Faegri, K., and Iversen, J., 1964, Textbook of pollen analysis: New York, Hafner Pub. Co., 237 p.

Florin, M. B., 1970, Late-glacial diatoms from Kirchner Marsh, southeastern Minnesota: Nova Hedwigia, v. 31, p. 667-756.

Fox, J. L., Odlaug, T. O., and Olson, T. A., 1969, The ecology of periphyton in western Lake Superior, Part I, Taxonomy and distribution: Minnesota Univ. Water Resources Research Center Bull. 14, 127 p.

Hagerty, F. H. (Commissioner of Immigration), 1889, The Territory of Dakota, an official statistical, historical, and political abstract: Aberdeen, S.D., Daily News Print, 311 p.

Haworth, E. Y., 1972, Diatom succession in a core from Pickerel Lake, northeastern South Dakota: Geol. Soc. America Bull., v. 83, p. 157-172.

Holland, R. E., 1968, Correlation of *Melosira* species with trophic conditions in Lake Michigan: Limnology and Oceanography, v. 13, p. 555-557.

——1969, Seasonal fluctuations of Lake Michigan diatoms: Limnology and Oceanography, v. 14, p. 423-436.

Hostetter, H. P., and Hoshaw, R. P., 1972, Asexual development patterns of the diatom *Stauroneis anceps* in culture: Jour. Phycology, v. 8, p. 289-296.

Hustedt, F., 1930, Bacillariophyta (Diatomeae), *in* Pasher, A., ed., Die Süsswasser-Flora Mittel-Europas: Jena, Germany, Gustav Fischer, v. 10, 466 p.

——1937-39, Systematische und okologische Untersuchungen uber die Diatomeenflora van Java, Bali, und Sumatra: Archiv Hydrobiologie Suppl. 15, p. 131-177, 187-295, 393-506, 790-836; Suppl. 16, p. 1-155, 274-394.

Hutchinson, G. E., 1967, A treatise on limnology, Vol. 2, Introduction to lake biology and the limnoplankton: New York, John Wiley & Sons, 1115 p.

Janssen, C. R., 1967, A postglacial pollen diagram from a small *Typha* swamp in northwestern Minnesota, interpreted from pollen indicators and surface samples: Ecol. Mon. v. 37, p. 145-172.

Jorgensen, E. K., 1948, Diatom communities in Danish lakes and ponds: Det Kongelige Danske Videnskabernes Selskab, Biologiske Skrifter, v. 5, no. 2, 140 p.

——1957, Diatom periodicity and silicon assimilation: Dansk Botanisk Archiv, v. 18, no. 1, 54 p.

Kilham, Peter, 1971, A hypothesis concerning silica and the freshwater planktonic diatoms: Limnology and Oceanography, v. 16, p. 10-18.

Kolbe, R. W., 1927, Zur Okologie, Morphologie, und Systematik der Brackwasser-Diatomeen: Pflanzenforsuchung, heft 7, p. 1-146.

——1932, Grundlinien einer algemeinen Okologie der Diatomeen: Berlin, Ergebnisse der Biologie, v. 8, p. 221-348.

Larson, W. C., 1961, Spray irrigation for the removal of nutrients in sewage treatment plant effluent as practiced in Detroit Lakes, Minnesota: Trans. 1960 Seminar: Algae and Metropolitan Wastes Seminar: Cincinnati 1960, U.S. Dept. Health, Education and Welfare, p. 125-129.

Mann, W. B., IV, and McBride, M. S., 1972, The hydrologic balance of Lake Sallie, Becker County, Minnesota: U.S. Geol. Survey Prof. Paper 800-D, p. 189-191.

McAndrews, J. H., 1968, Pollen evidence for the protohistoric development of the "Big Woods" in Minnesota (U.S.A.): Rev. Palaeobotany and Palynology, v. 7, p. 201-211.

McBride, M. S., 1972, Hydrology of Lake Sallie, northeastern Minnesota, with special attention to groundwater-surface interaction [M.S. thesis]: Minneapolis, Univ. Minnesota, 62 p.

McClure, P. F., 1887, Resources of Dakota: Sioux Falls, S.D., Dept. Immigration and Statistics, 498 p.

Megard, R. O., 1969, Algae and photosynthesis in Shagawa Lake, Minnesota: Minnesota Univ. Limnol. Research Center Interim Rept. 5, 20 p.

——1970, Lake Minnetonka: Nutrients, nutrient abatement, and the photosynthetic system of the phytoplankton: Minnesota Univ. Limnol. Research Center Interim Rept. 7, 210 p.

——1972, Phytoplankton, photosynthesis and phosphorus in Lake Minnetonka, Minnesota: Limnology and Oceanography, v. 17, p. 68-87.

Merilainen, J., 1971, The recent sedimentation of diatom frustules in four meromictic lakes: Annales Bot. Fennica, v. 8, p. 160-176.

Minnesota Pollution Control Agency, 1972, Wastewater disposal facilities inventory, January 1, 1972: Div. Water Quality, Public Works Sec., 66 p.

Moyle, J. B., 1954, Some aspects of the chemistry of Minnesota surface waters as related to game and fish management: Minnesota Dept. Conservation Inv. Rept. 151, 36 p.

Neel, J. K., Peterson, S. A., and Smith, W. L., 1973, Weed harvest and lake nutrient dynamics: U.S. Environmental Protection Agency, Ecol. Res. Ser., EPA-660/3-73-001, 91 p.

Patrick, R., and Reimer, C. W., 1966, The diatoms of the United States, Vol. 1: Acad. Nat. Sci. Philadelphia Mon. 13, 688 p.
Schaumburg, F. D., 1973, The influence of log handling on water quality: Washington, D.C., U.S. Environmental Protection Agency, Environmental Protection Tech. Ser., EPA-R2-73-085, 105 p.
Schindler, D. W., and Holmgren, S. K., 1971, Primary production and phytoplankton in the Experimental Lakes area, northwestern Ontario, and other low-carbonate waters, and a liquid scintillation method for determining ^{14}C activity in photosynthesis: Fisheries Research Board Canada Jour., v. 28, p. 189-201.
Schultz, M. E., 1971, Salinity-related polymorphism in the brackish-water diatom *Cyclotella cryptica:* Canadian Jour. Botany, v. 49, p. 1285-1289.
Stark, D. M., 1971, A paleolimnological study of Elk Lake in Itasca State Park, Clearwater County, Minnesota [Ph.D. thesis]: Minneapolis, Univ. Minnesota, 178 p.
Stein, G. P., 1971, Lake Minnetonka, historical perceptions of an urban lake [Ph.D. thesis]: Minneapolis, Univ. Minnesota, 241 p.
Stockner, J. G., 1971, Preliminary characterization of lakes in the Experimental Lakes area, northwestern Ontario, using diatom occurrences in sediments: Fisheries Research Board Canada Jour., v. 28, p. 265-275.
Stockner, J. G., and Armstrong, F. A., 1971, Periphyton of the Experimental Lakes area, northwestern Ontario: Fisheries Research Board Canada Jour., v. 28, p. 215-229.
Stoermer, E. F., and Yang, J. J., 1970, Distribution and relative abundance of dominant plankton diatoms in Lake Michigan: Univ. Michigan Great Lakes Research Div., Pub. 16, 64 p.
Stoermer, E. F., Bowman, M. M., Kingston, J. C., and Schaedel, A. L., 1974, Phytoplankton composition and abundance in Lake Ontario during IFYGL: Univ. Michigan Great Lakes Research Div. Spec. Rept. 53, 373 p.
Swain, A. M., 1973, A history of fire and vegetation in northern Minnesota as recorded in lake sediments: Quaternary Research, v. 3, p. 383-396.
Tarapchak, S. J., 1972, Studies of phytoplankton distribution and indicators of trophic state in Minnesota lakes [Ph.D. thesis]: Minneapolis, Univ. Minnesota, 204 p.
Tippett, R., 1964, An investigation into the nature of the layering of deep-water sediments in two eastern Ontario lakes: Canadian Jour. Botany, v. 42, p. 1693-1709.
Waddington, J.C.B., 1969, A stratigraphic record of the pollen influx to a lake in the Big Woods of Minnesota: Geol. Soc. America Spec. Paper 123, p. 263-281.
Watts, W. A., and Bright, R. C., 1968, Pollen, seed, and mollusk analysis of a sediment core from Pickerel Lake, Day County, South Dakota: Geol. Soc. America Bull., v. 79, p. 855-876.
Weaver, J. E., and Clements, F. E., 1929, Plant ecology: New York, McGraw-Hill Book Co., 520 p.
Wilcox, A. H., 1907, A pioneer history of Becker County, Minnesota, including a brief account of its natural history: St. Paul, Pioneer Press Co., 757 p.
Will, G. F., 1946, Tree ring studies in North Dakota: North Dakota Agriculture Expt. Sta. Bull. 338, 24 p.
Wodehouse, R. P., 1959, Pollen grains: New York, Hafner Pub. Co., 574 p.
Wright, H. E., Jr. 1968, History of the prairie peninsula, *in* The Quaternary of Illinois: Urbana, Ill., Univ. Illinois, Coll. Agriculture Spec. Pub. 14, p. 78-88.
Wright, H. E., Jr., and Rhue, R. V., 1965, Glaciation of Minnesota and Iowa, *in* Wright, H. E., Jr. and Frey, D. G., eds., The Quaternary of the United States: Princeton, N.J., Princeton Univ. Press, p. 29-41.
Wright, H. E., Cushing, E. J., and Livingstone, D. A., 1965, Coring devices for lake sediment,

in Kummel, B., and Raup, H. M., eds., Handbook of paleontological techniques: San Francisco, W. H. Freeman & Co., Pubs., p. 494–520.

MANUSCRIPT RECEIVED BY THE SOCIETY MAY 6, 1974
REVISED MANUSCRIPT RECEIVED MARCH 11, 1975
MANUSCRIPT ACCEPTED APRIL 9, 1975
CONTRIBUTION NO. 131, LIMNOLOGICAL RESEARCH CENTER, UNIVERSITY OF MINNESOTA
AUTHOR'S PRESENT ADDRESS: U.S. GEOLOGICAL SURVEY, FEDERAL CENTER, DENVER, COLORADO 80225